TEACH YOU

D0618933

UNDERSTANDING STATISTICS

This book examines how statistics are collected and presented, and how the ordinary man or woman can learn to distinguish the facts from the fallacies.

In the press and on radio and television, statistics are continually quoted as establishing the true facts about a situation —about crime, politics, the economy or even the 'hit parade'. But frequently statistics are misrepresented or are inaccurate and there is a temptation on the part of experts to misinterpret statistics in order to further their arguments.

This book—clearly, concisely and without the use of complicated mathematics—helps the reader to understand statistics and to refute anyone who confidently states that 'the statistics show beyond doubt that . . .'.

TEACH YOURSELF BOOKS

UNDERSTANDING STATISTICS

J. M. Leonard, M.A.

ST PAUL'S HOUSE WARWICK LANE LONDON EC4P 4AH

First printed 1974

Copyright © 1974
The English Universities Press Ltd.

ISBN 0 340 18259 8

Printed in Great Britain
for The English Universities Press, Ltd.,
by Richard Clay (The Chaucer Press), Ltd., Bungay, Suffolk

Contents

Preface

An understanding of the nature of statistical reasoning is becoming increasingly important for the ordinary citizen. No doubt there are many who would wish to question this statement, but perhaps they are not among those with sufficient interest in the subject to have picked up a book on it. The aim of this book is to assist those who wish to develop such an understanding and, at the same time, to clarify some of the sources of confusion with which the subject is beset.

Mathematical analysis has been rigidly excluded, and nothing more difficult than multiplication and division, percentages, and the occasional square root is required at all. There are a few chapters which use some relatively difficult arithmetic, but readers not wishing to check every detail of the calculations will still be able to follow the general line of argument.

Statistics is a subject whose different aspects are interrelated in a fairly complex way. Anyone writing about it therefore has to make a frequent choice between repeating a point mentioned elsewhere and giving a cross-reference. My aim has been to give cross-references for major items, but to allow some repetition of minor ones, in the hope that the reader will not find too much of either. A related point is that I do not expect all readers to cover every page: many will find that Chapters 3, 4 and 5 contain mostly familiar material, for example. Chapters 11 to 16 cover a few of the many subjects in which statistical analysis is of fundamental importance, and can be read independently of each other; I expect that most readers will choose from this section only the parts which interest them.

I am grateful to Alan Cooper and Peter Pulford for their most helpful comments on the draft. My thanks also go to

Miss A. M. Walker of the British Market Research Bureau Limited for information about the 'Top of the Pops' record chart, to Mr G. Whitaker for his comments on the draft of Chapter 13 and, most particularly, to Dr E. A. Wrigley for his expert corrections of an amateur demographic statistician's mistakes in the draft of Chapter 16. I am grateful also to Frank Sharman for his help with the computer calculations for the population models used in this chapter.

Any writer who sets out to expose some of the errors in other people's statistics thereby invites his readers to search out the mistakes in his own. It is too much to hope that they will all be unsuccessful in their search, and I would ask the successful ones to point out the errors they find, so that corrections can be made. In a book which draws its illustrations from a wide range of subjects, including some of which I have no specialised knowledge, it is possible that one or two errors of a more technical nature have crept in and, once again, I should be grateful to readers who were able to draw my attention to them.

J.M.L.

1 Introduction

One of the aims of education in its widest sense has always been that a man should be literate or, in other words, that he should have a good command of his native language. Another, likewise, has been that he should be skilled in the use of numerical concepts or, as it is nowadays said, numerate. Often this is interpreted as having the ability to solve arithmetical problems connected with income tax, the fuel consumption of a car or economy in shopping, together with some understanding of algebraic and geometrical methods. But if we, as numerate citizens, are to be able to interpret intelligently the numerical information which assails us each day, something more is required—something which, incidentally, finds little place in the curriculum of most schools.

Suppose, for example, that we are attempting to follow the progress of some wage negotiations in a large firm. A wage demand has been made, because of rising prices and increased productivity, and to gain parity with wage rates in a competing firm. Negotiations are proving difficult and a strike is threatened. Where does our sympathy lie?

First, the rising prices. How much have they risen? This requires, typically, the interpretation of some kind of retail price index, and is by no means a matter of simple arithmetic. Is the index the right measure for the change? It might be misleading because it was based on an out-of-date expenditure pattern. Also, the calculated rise must be compared with any rise there may have been in wages over the same period, and this may be made to look more favourable to one side or the other by choosing, say, basic rates rather than actual wages, or rates for the lowest-paid workers rather than average rates. And what happens when the nature of a particular job has changed, perhaps as a result of mechanisation? The problem begins to look formidable even before we start to look at either productivity or parity, each of which poses much greater difficulties.

Usually, of course, the detailed information required to reach a fair conclusion is not available to the outsider. His newspaper will quote (at best) one or two examples of wage rates as given by one side or the other, some figures for the second firm which may or may not be directly comparable, some figures for productivity on an undefined basis of calculation, and a percentage figure for the cost-of-living increase. Since those involved in the dispute have much more information than this and still cannot reach agreement, the outsider can hardly hope to do so. But the essential point is that we should be able to assess critically the information, however inadequate, which is available to us, and to recognise the fallacies which often lie behind the figures quoted. If we are to be considered numerate, we need to be trained in the interpretation of statistical data, and that is what this book is all about.

After this introductory chapter, there is a brief account of some of the principal sources of statistical information which are readily available through libraries. The next three chapters describe the ways in which such information is analysed and prepared for the reader. Then there is a section on probability and its practical uses in statistics, particularly in drawing conclusions from samples, followed by a chapter on statistical experiments. The next chapter provides the central theme of the book, with the title 'How to read a statistical argument', and this is followed by a short introduction to some further statistical techniques. The remaining six chapters describe a few examples of statistics in use, chosen from a wide range of possible subjects partly on account of their inherent interest and partly to show that statistics is not just for statisticians but is also involved in many different kinds of job, and in many different aspects of the life of the ordinary citizen.

One point which should be mentioned here is that many concepts used in everyday speech are statistical in nature, even though they include no explicit numerical information. For example, 'public opinion' is a statistical concept, and the whole basis of weather forecasting is the interpretation of large quantities of numerical information. A carcinogenic substance is one which tends to encourage the onset of cancer, but in a manner which is not predictable for each individual case and has to be expressed statistically. The point at which something ceases to be 'normal' and becomes 'abnormal' can often only be expressed in statistical language. And words such as *probable, likely, almost, many, often, some, most, usually* and *sometimes* are used generally in statistical statements. The last two sentences are themselves examples of this. In common with matters which are more overtly statistical, such terms and statements should be interpreted with care: the reader should look out for them.

The growth of statistics

The collection and study of numerical data is very much older than the word 'statistics' itself, which dates only from the late eighteenth century. In a sense it is as old as counting, which was used in the earliest civilisations—Egyptian, Sumerian, Chinese, Greek and Roman. One early example is the census of Israel taken by Moses during the Exodus, which is recorded in the first four chapters of the Book of Numbers.

The Incas of Peru had a highly efficient system for collecting and recording data about every aspect of their national life, interesting both because it probably developed independently of the civilisations of the Old World and because the methods employed were as sophisticated as any in use before the invention of electronic data-processing.

The best-known example in English history is the Domesday Book of 1086, which recorded information about population, occupations, land values and use, housing and live-stock. The survey was so thoroughly done, the chronicler wrote, that there was 'not one yard of land, nor one ox nor one cow nor one swine left out'. Since the purpose of the survey was to raise taxes, this must be regarded as a somewhat optimistic conclusion.

In modern times the growth of statistics may be traced back to the early years of the Royal Society in the 1660s, when John Graunt, William Petty and others began to think about statistics in social terms—births and deaths, family structure, housing, diseases and the supply of food. Records of births (initially baptisms), marriages and deaths began to be kept more systematically, although this did not become a legal requirement in England until 1837; records were incomplete until about 1870.

During the nineteenth century the types and amount of data available increased rapidly. Statistics of manufacturing and trade, of taxation and wealth, of geography and weather, of crimes and diseases, as well as population figures—all these became more and more widely kept and published. For the most part Government departments were responsible, as the policy-makers came to realise that the efficient collection and tabulation of numerical data was the first step towards efficient administration. During this century the process has accelerated still further, and part of the purpose of this book is to show something of the present scope of the subject.

Probability methods

In parallel with this growth in the use of statistics, and to a large extent independently of it, there grew also the study of probability. Men have played games of chance for as long as recorded history, and six-sided dice over 4000 years old have been found; it is reasonable to assume that an empirical understanding of probability developed at the same time. One of the first to write about it was the astronomer Galileo, in about 1620, and the subject came into greater prominence in 1654 when a gamblers' dispute was resolved in correspondence between two of the great mathematicians of the time, Pascal and Fermat. It continues now as a branch of pure mathematics in its own right, but it is also fundamentally involved in the analysis of statistical data.

The first notable point at which probability and statistics came together was the publication in 1693 of the Life Table of Edmund Halley, better known for his work on comets. This was based on the records of births and deaths in the city of Breslau (modern Wroclaw, in Poland), and its aim was 'to ascertain the price of annuities upon lives'. For the first time the growing practice of life assurance was placed on a sound theoretical basis.

A further landmark was de Moivre's development of the laws of probability in the case of experiments repeated a large number of times, culminating in the discovery in 1733 of what is now called the Normal Distribution. The mathematical theory of statistics (mentioned only briefly in these pages) may be traced back to this time.

Another important contact between probability and statistics was Mendel's work on heredity, published in 1866. In his best-known experiment, Mendel crossed two pure-bred varieties of sweet pea, one with yellow seeds and one with green, producing a hybrid with yellow seeds. The hybrids, when self-fertilised, produced offspring which were yellow and green in the ratio of almost exactly 3:1, out of about 8000 plants. The balance between the types of genes on a probability basis suggested just this value.

In the early years of this century these two lines of progress began to fuse more completely. Of many names associated with the rapid growth during these years, one stands out—that of Sir Ronald Fisher, Director of the Agricultural Research Station at Rothamsted in Hertfordshire. His book *Statistical Methods for Research Workers*, published in 1925, is a classic, and the statistical tables of Fisher and Yates are still the best-known set available.

But, although the distinction between descriptive statistics and probability methods is no longer such a clear one, it will still be helpful to keep it in mind as we start to look at the range of the subject today. Most of the sources of statistics mentioned in Chapter 2 give information which may be understood without reference to probability concepts; the same is true of the methods of presentation and analysis described in Chapters 3 to 5. But the sampling methods introduced in Chapter 7 depend wholly on the concept of probability. In later chapters the reader will be able to use his understanding of the distinction to see for himself where probability methods are being used, and likewise as he reads a statistical argument in his newspaper. It is an important distinction, and one which is far too little understood.

2 Sources of statistical information

The extent of the information now available is immense, and even a section of it which covers one subject in a single country is in many cases too big for an individual to assimilate completely.

The major sources of national data are the statistical departments of each government concerned. In most countries there is a specialist department whose sole purpose is the collection and publication of this information, but in the United Kingdom the system is decentralised. Every department of government has its own statistical section. The Central Statistical Office, which is attached to the Cabinet Office, is responsible for co-ordination, and the head of this department is also designated Head of the Government Statistical Service.

As a small indication of the range of Government statistics, the list of regular publications issued by the CSO covers about 280 items, from small pamphlets to the five volumes of *Annual Statistics of Overseas Trade* costing a total of over £40. In addition, there is a wide variety of occasional publications, many of them statistical in nature. Most departments publish an annual report, supplemented in some cases by monthly and weekly bulletins giving the most up-to-date information. The principal items of general interest are collected together and published in various single volumes, of which the *Annual Abstract of Statistics* and the *Monthly Digest of Statistics* are the most comprehensive.

Population statistics are based partly on the decennial Census and partly on the routine recording of births and deaths, combined with the available information on migration. The results of the Census are published in a large number of volumes which give information on several matters in addition to population: household composition, internal migration, qualified manpower, housing, transport and places of work. A partial census is also taken at the half-way point in each ten-year period.

The Registrar General's annual report gives information about births, marriages and deaths, stillbirths and abortions, immigration and emigration, notifiable diseases and causes of death. All the figures are analysed in one or more different ways—by age, month of the year and so on. An annual mid-year estimate of the total population is issued, with analysis by local government areas. From all these figures, estimates are given annually of mortality rates and expectation of life, in a form known as a Life Table; more detailed versions of these are published at ten-year intervals, giving rates for various occupations. Population projections for about forty years ahead are also published annually. (Demographic statistics are covered in more detail in Chapter 16.)

Social statistics from a wide range of sources are combined in a useful volume called *Social Trends*, which has been published annually since 1970. This was begun as part of a clear and commendable policy of making the Government Statistical Service more widely known and its publications more widely understood and used. *Social Trends* covers much of the same ground as the *Annual Abstract*, but is rather easier to read because it employs visual presentation methods as well as tabulated figures. Topics in it include the health and welfare services, housing, education, crime and punishment, prices and incomes, road accidents, population and related statistics. A similar though smaller volume called *Economic Trends* is published monthly.

Changes in some price and wage levels are recorded by means of an index, a device discussed in Chapter 4. The best-known of these is the Index of Retail Prices, but there are others covering matters such as prices of new houses, imports and exports by value, and certain commodity prices.

The information needed to calculate the Index of Retail

Prices is derived mainly from a continuous Family Expenditure Survey, begun in 1957. This collects information from 16 families in each of 168 carefully balanced areas, 4 times a year—a total of 10 752 families. The family is asked to supply details of income from all sources, and information about expenditure outside the normal weekly pattern. Then each member of the family over the age of 15 is given a diary in which to record every sum of money spent, for two 7-day periods. A nominal payment of £2 is made to each person who takes part over the whole period: this acts as a small encouragement to the families selected, as there is no compulsion. (The census is the only enquiry of this kind in which participation is legally enforced.) The response rate is about 69%, which is a good deal less than that generally agreed to be necessary for an accurate survey. Also, there is evidence of under-reporting of expenditure in certain categories, such as tobacco, meals outside the home and, above all, alcohol.

The FES is only one of a wide range of surveys, both continuous and occasional, which are carried out by the Social Survey Division of the Office of Population Censuses and Surveys. The other division deals both with the census itself and with the registration of vital statistics—the word 'vital' being used here in its correct sense of referring to birth, life and death. The two departments merged in 1970, and the different sides of the work are now seen as complementary in a way that was not always evident in earlier years.

One result of this co-operation has been the creation of the General Household Survey, which is again a continuous survey with a sampling arrangement similar to that of the FES. The contact with each family consists of an interview lasting about an hour, with questions covering family composition, housing and home amenities, employment, travel, income, illnesses, contacts with welfare services and changes of address in the previous five years with reasons for each change. The response rate is about 82%. Much of the information obtained in the GHS is similar to that derived from the census, but this does not imply wasteful duplication—in fact, the two forms of enquiry complement each other most effectively.

A further survey, based on the existing written records rather than on personal contact, is also being started. It is a cohort study: the progress of a group of people, about 1% of the whole population, is to be followed over a period of years, using information drawn from census and registration data and from the central register of the National Health Service. As the survey continues, one baby of every hundred born, and likewise one immigrant in every hundred, will be added to the group. The primary purpose is to trace possible connections between certain diseases, such as cancer, and the social and educational background of those who contract them.

The chief merit of a census, compared with surveys such as these, is the obvious one of completeness, which means that more information can be derived about variations from area to area. Figures for minority groups are also more accurate, although it is not by any means always true that census figures are better than survey ones. The element of legal compulsion in the census also gives it an advantage in the form of a higher response rate, but the proportion of errors (deliberate or otherwise) is more difficult to judge.

The advantages of the survey derive from its much smaller size. It is faster, and so gives more up-to-date results at the time of publication; it can be carried out more frequently so that, by the time one set of results is starting to get stale, a newer set becomes available. Taken together, these two points counteract the main disadvantage of the census, which is that, allowing for the two years or so required to publish the results, the figures may be up to twelve years out of date before the next set are published.

Also, a survey can be carried out by professional interviewers, so that error rates are reduced; the extent of the possible errors is also easier to estimate, even if on a rather subjective basis. The details of the survey can be modified to take account of experience, whereas a poorly worded census question (that in 1971 on live-born legitimate children was one example) can mean an irretrievable loss. And, finally, a difference more important than any of these—a survey can be carried out at greater depth.

As an example of this last point, the census asks for information about changes of address within the previous five years. But questions about the *reasons* for these changes can only be asked by an interviewer who has gained the respondent's co-operation. Details of illnesses, also, can only be found by verbal questioning if the results are to be at all accurate, because of the difficulty of defining the word 'illness' in a simple enough way. It is possible that income details will be stated more accurately in an interview than on a form, although only if the respondent is convinced that what he says will remain confidential. It is an important part of the interviewer's job to explain this.

Understandably, some people are hard to convince. It seems inherently unlikely that one Government department would be able to guarantee that none of the information given to it would be passed to another department, particularly when the Tax or Social Security departments might be involved. But, even when no explicit undertaking has been given, such transfers of information relating to particular individuals require the authority of Parliament. It is understood on all sides that much of the information collected by census and survey would not have been revealed by those concerned except under guarantee of secrecy, so that any breach of the confidence rule would in the long term be self-defeating as well as dishonest. Although reported breaches are very rare, the use of computers, not only for the processing of information but also for its transfer from place to place, means that the matter will require constant vigilance. In the cohort study mentioned earlier, those involved are to be identified by a code number only, so that information relating to named individuals is no less secure than it was when the survey started.

The other surveys carried out by the Social Survey Division cover many different subjects, sometimes chosen by the department itself and sometimes at the request of other sections of the Government. Three examples will be sufficient to mention here. A selection of the families already contacted in the GHS agreed to co-operate in a survey of people who had received hospital treatment—a good example of how one investigation can reveal a need, not previously recognised, which can then be followed up by a special study. A Family Intentions Survey was undertaken to assist in forecasting trends in the birth-rate pattern—notoriously, one of the most difficult problems the demographic statistician has to tackle. No doubt the stated intentions will not all be fulfilled in practice but, if the discrepancy is not too great or if it is the same as that shown by later surveys of the same type, the results will be better than inspired guesswork. Thirdly, a survey of the handicapped was carried out—a particularly difficult kind of work, partly because of the small proportion of handicapped in the whole population, partly because the nature of the group meant that in any general survey a low response rate was to be expected from them. The survey showed, incidentally, that there were over $1\frac{1}{4}$ million handicapped adults in England and Wales, of whom 1 in 5 lived alone: these results, and the more detailed ones for particular areas, should prove immensely useful to those with responsibilities towards the handicapped and so, in turn, to the handicapped themselves.

The Census of Population is of course well known, but there are two other censuses also—those of Production and Distribution. Each of these is comparable with the Population Census in the magnitude of the task: fewer units are covered, but each one involves much lengthier and more detailed work than a family census form, both for the firms concerned and for the census officials. The different stages in the preparation of the Census of Production give some indication of what is involved.

As the completed forms come in, they are edited by hand as a rough check on completeness and then coded, including the difficult job of classifying by industry, for input to the computer. The computer carries out what are called credibility checks, by working out the values of such measurements as the mean output per head, mean salary of employees and so on, and pointing out any which fall outside certain pre-set limits. This helps to trace the mistakes in form-filling which are inevitable in such a large-scale operation. Any missing or incomplete data are estimated or imputed, with checks back to the firm where necessary. The figures are then aggregated, and checked before publication to ensure that no figures relating to a particular firm can be deduced from the totals, to ensure compliance with the confidence rule. While this is in progress, firms which are slow to send in their returns are contacted by letter and telephone to spur them on; even so, some take over twelve months.

A post-census survey is now standard practice for all three censuses. Asking useful questions of industrial and commercial firms demands considerable care, and often much specialised knowledge of accounting methods, technical classification and so on. The purpose of a follow-up survey is therefore twofold: to check the accuracy of the answers which have been given, by asking more detailed questions than is possible in a full census, and to find out if the right questions were asked in the first place. (In the case of the population census, the purpose is more limited, and the procedure here could well be improved.) Any defects which show themselves can then be put right in the next census.

For similar reasons, a pilot survey is normally carried out before a full census, so that faults in the questions can (with luck) be eliminated (see also Chapter 7). A further advantage of surveys of this kind is that they assist both surveyors and surveyed to understand each other's problems, because of the personal contact involved.

Complete response in a census is an ideal never reached, whether the methods of persuasion are legal threats or polite reminders. In every case, therefore, a choice has to be made between publishing incomplete figures quickly and waiting for better ones, which even then will still not be perfect. The usual method is to make speed the principal aim, and to issue revised figures later when these become available. This sometimes causes annoyance to those who use the figures but, short of a rapid improvement in the standards of human punctuality and reliability, no other method is feasible. The actual processing of the figures is now much faster than it was, as a result of using computers, but this does not get over the problem of the firm which sends its returns in late. In the Census of Population, an individual has the right to send his form direct to the Census Office rather than hand it to the local enumerator, and in 1971 so many people did so that the entire operation was delayed for about two months.

This is not the place to give a full account of all the Government statistics available—in fact, several major subjects have barely received a mention, particularly in the business and financial fields. The best way to find out more is to look at a copy of the *Annual Abstract*, which includes references to other sources for more detailed information.

Most countries publish their own statistics on similar lines, with varying degrees of completeness and accuracy. National statistics are also collated and published by the United Nations Statistical Office in an annual volume with a monthly supplement, and in various specialised publications such as the *Demographic Yearbook*. Agencies such as

UNESCO and the FAO publish statistical reports. The accuracy of UN publications is of course only as good as that of the source material, but commentaries on the accuracy of the data are often included: for example, the population figures are given a rating from A for a full survey to D for 'conjecture'. The European Economic Community also has a statistical department which publishes reports on similar lines to those of the UN.

Other organisations which publish statistical reports are many and various. Local government departments, trade associations, professional organisations and journals, and trade unions publish information about their own spheres of activity, and many firms are large enough for their annual reports to be of general statistical interest. University departments and independent research bodies, often financed by industrial firms or charitable trusts, also publish their reports, as do the various Government research bodies; only a small fraction of these ever find even a brief mention in the newspapers. A few newspapers and magazines carry out statistical studies themselves, though more often they commission one of the independent survey organisations such as National Opinion Polls or the Opinion Research Centre to do the work. There is a surprisingly large number of these specialist survey firms, which are mostly engaged on market research work—a major use of statistical methods, though one which stays mostly out of the public eye.

Television and radio audience research is carried out both by the BBC and by a specialist firm commissioned by an organisation known as JICTAR, the Joint Industry Committee for Television Advertising Research, which covers both BBC and IBA. The BBC's own research is designed simply to test reactions to their own programmes, but for commercial television and radio the figures are required by JICTAR in order to fix advertising rates, which vary according to the numbers watching or listening at any given time.

Not only the totals are required, but also the numbers in each age group and income group. Some of this information is released to the Press as being of general public interest, and similarly for the BBC's own figures. Research to estimate readership figures for newspapers and magazines is carried out on the same lines as that for commercial broadcasting, and for similar reasons.

Consumer publications such as *Which?* provide information designed to assist their readers to spend their money wisely—one of the most useful statistical sources for the ordinary citizen. But the most frequently used source for most people will remain their daily newspapers, and many of the examples used in the following chapters have been chosen with this point in mind.

Useful summaries of statistical information are given in such publications as *Whitaker's Almanack*, the *Statesman's Year Book*, and the annual Stationery Office guide, *Britain, an official handbook*. Two 'popular' versions of mainly Government statistics have been published, both by Penguin Books: *Britain in Figures* and *Facts in Focus*. In terms of value for money, these are as good as anything available; *Social Trends* is much better but much more expensive.

The range of coverage by all these sources is very wide, as no doubt will be quite clear to the reader. And the range of quality is wide also. The most obvious defects, both of omission and commission, are to be found (predictably enough) in the statistics produced by the small pressure group rather than in those of the big Government department. But this is by no means always the case.

In spite of this wide coverage, there remain many gaps, and the nature of many of them is such that they are not likely to be filled. It may be of interest to mention some of them.

National security is perhaps the most obvious reason: the information service of the Central Statistical Office is not likely to give much away about the number of espionage agents in Government service. Some countries interpret the limits of national security rather widely. Commercial and financial security is a reason for another big gap—it is well known that an annual balance sheet is designed as much to conceal information as to impart it. Even the information which registered companies are legally required to supply is far from complete, and the problem of keeping the Companies Register up to date has so far proved insuperable. For businesses other than registered companies, not even a list of names exists: those responsible for the 1971 Census of Distribution had to construct a list of retailers partly by physically looking for them. But work on such a list is now in progress.

Less obviously, information about the state of health of the whole population is very limited, because preventive medicine and dentistry occupies such a small part of the Health Service's work. There is immense scope for filling in some of the gaps here, and using the results to increase the proportion of early diagnosis of a wide range of medical and mental conditions.

There are some minority groups who are likely to find themselves left out, or in the non-respondent category, in even the most carefully conducted census or survey: the vagrants and gypsies, the illegal immigrants, the small tribal group in an underdeveloped country. (A survey of vagrants in some of the principal cities in England has been carried out, but there are only rough guesses about numbers in other areas.) In the same category comes information which people wish to hide, on, for example, the use of drugs or illegal activities which do not affect other people directly. Even where other people are affected, particularly in sexual offences, the number of unrecorded crimes is sometimes very high. (A few of the many problems concerning criminal statistics are covered in Chapter 15.)

Statistics of people or things which cannot be adequately

defined are major gaps in our knowledge. The most obvious example concerns religious adherence, on which the available information is often of doubtful value and sometimes quite meaningless. Membership of churches is in most cases undefined, or defined in such a way as to be unrelated to attendance at worship or to any other reasonable measure of true religious activity—if indeed this is measurable at all. Information about actual attendances at worship is not usually available, or is based on a few particular occasions when attendance is known to be high. Also more or less undefinable, though for very different reasons, is personal wealth, and, at the highest levels, the same might be said about income. At the other end of the scale, standards of poverty are very difficult to define, though the available methods have improved a great deal since the pioneer work of Seebohm Rowntree in 1901.

Next, there are various things which it would be quite interesting to know but which it would take far too much trouble and expense to find out. A full statistical picture of the national transport system, for example, would be of immense value. A survey designed to produce such a picture would be too complicated to carry out; simplification would cause too many errors for the results to be useful. Even the first stage of the work, the design of a sampling system, would be almost impossibly difficult, and the seasonal variations in the traffic pattern would also cause problems.

Some kinds of survey question cannot be expected to yield any useful information, even if they can be asked at all: on standards of personal honesty, private habits and thoughts. You can make a person answer a question by force of law, but you cannot make him give a true answer. The under-reporting of alcohol consumption in the Family Expenditure Survey which was mentioned earlier is one example; another is the series of Kinsey reports, whose results are open to serious doubts on these grounds.

There are also some subjects which are not particularly private but on which it would be difficult to ask questions in any useful way: job satisfaction, personal friendships and matters concerning motives. The problem here is that the individual normally responds subjectively to these things, but the researcher must try to look at them objectively if he is to make any sense of the results. Since this can very rarely be done the consequence is, in every case, a gap in our statistical picture of the community.

There are also questions which cannot be asked because they are politically too sensitive—questions on racial background and skin colour being obvious examples. Similarly, there are limitations to the questions which may be asked of children, so that no statistical survey can tell us what really goes on in the classroom, and (for this and many other reasons) debates about the relative merits of different styles of teaching are likely to be inconclusive. It may be no co-incidence that three matters which are the cause of continuing public controversy are also matters which are difficult to assess statistically: changing patterns in education, the consequences of having different racial groups in a single community, and immigration.

Finally, there are matters which are ephemeral. It would revolutionise the science of market research if some way could be found of testing the reaction of a suitable sample of viewers to a particular advertisement which they were seeing on the television screen. The simple truth is that the effect of a single advertisement, either in single or repeated showings, is unknown, and only the changes in sales figures give any idea at all. The individual reactions to it can be neither quantified nor remembered for the benefit of researchers a few days later. Perhaps it is not one of the most important gaps in our knowledge, but it is a very real one.

3 A first look at the figures

The first contacts which most of us had with arithmetic probably led us to think of figures as straightforward, obedient objects which gave unarguable results when added, multiplied and so on, provided that we followed certain agreed rules. For figures in an abstract context, as in the classroom, this is of course quite true. But in real life it is not so.

The basic information of statistics consists of figures which are counted, measured or estimated. But we can never be sure, except when counting fairly small numbers, that the stated values are correct. In statistical work a difference between the stated value of a figure and its true value is called an **error**—not implying a mistake, though there may well have been one. Thus, if a distance is stated to be 17 km when a more exact value is 17·189 km, the error is about 0·189 km. Even 17·189 is not fully exact, of course, and so the error is not exact either.

A quantity which is counted or measured is a **variable**. When measurement, as opposed to counting, is used, there is no absolute limit to the degree of accuracy which can be obtained—in other words, total exactness of a measured quantity is impossible. The practical limit may be a very high one, as in cases such as times, distances and angles when accuracy of the order of one part in a hundred million is possible with high-quality instruments. But, more often, limits such as the nearest millimetre or hundredth of a second are used.

For practical purposes, a scale of measurement goes in steps, just as in a counted scale or a currency scale. For example, times may be measured in seconds or hundredths of a second, and a count is made of the number of complete units. A variable restricted to countable steps is **discrete**. In theory this restriction can be removed, giving a variable which is **continuous**. Continuous variables are really a part of theoretical rather than applied statistics, but it is worth noting that when the discrete steps are small enough the variable may look as though it is continuous.

The first stage (apart from mistakes) at which an error may be introduced is in choosing a point on the selected discrete scale to fit the measurement actually made. This is partly because there are two distinct ways of doing it and partly because of the problem of marginal cases.

Suppose, first, that we are measuring ages, and that the discrete scale to be used is that of the whole numbers of years. We can either ignore fractions of years, so that a man aged 23 years 10 months is counted as 23, or we can take the nearest year, so that he is assumed to be 24. There are advantages and disadvantages in both, but, whichever is used, the choice must be stated. One difference may be illustrated thus: if 10 000 babies were born each year the number who would be recorded in a census as having age 0 (assuming for simplicity that none died) would be 5000 on one method and 10 000 on the other.

A further difference is as follows. If ages are measured to the nearest year, all those between $23\frac{1}{2}$ and $24\frac{1}{2}$ will be counted as 24. In a large group of assorted ages, therefore, some will be overestimated by a few months and others correspondingly underestimated. For certain purposes—such as finding the average age of the group as a whole—the errors will thus tend to cancel each other out. Such errors are known as **unbiased** or **compensating** errors.

By contrast, the errors introduced by taking the age to be the number of completed years will be **biased** or **cumulative** errors, because every age will be underestimated by an amount up to just less than a year. If the ages are now used to work out an average the errors will not tend to cancel each other out and the result will be an underestimate, to the extent of about six months.

This does not imply that the second method should not be used, for in turn there are other reasons why it is more convenient for certain purposes. But when it is used, allowance for the bias must be made. Rounding off 'to the nearest' is in practice the commoner method.

Suppose now that we are measuring time to the nearest second. Then a time which is actually 13·49 seconds will count as 13, and 13·51 is taken to be 14. But a stop-watch reading to a tenth of a second cannot distinguish between 13·49 and 13·51. Errors in marginal cases such as this are unavoidable, and can only be reduced by using measuring techniques of much greater accuracy than that desired for the final figures, which is clearly uneconomic. They are, however, always small and rarely a cause of serious difficulty.

There are several ways of dealing with them when they do arise. First, one can look back to the source from which the doubtful figure came and either remeasure it or calculate it more accurately. This is clearly foolproof when it can be done. The next best is to show a marginal figure in the form 4·862(5), the 5 not being an accurate digit in the sense that the others are. But this is inconvenient in tabulation. If a choice has to be made between rounding up and rounding down, it is best to do half one way and half the other so that errors tend to cancel out: the obvious way is to do them alternately, but it is better to round off so that the rounded digit of the result is even. If division by two is later required, no further ambiguity then arises. Finally, the rounding may be done so that the error, if any, is on the safe side: if a manufactured article is to be passed by an inspector if its length, to the nearest 0·1 mm, is in the range 75·8 mm to 76·2 mm, then an item measured as 76·2(5) should be rejected.

One consequence of rounding is that the total of the

figures in a column may appear to be incorrect. Suppose, for example, that a quantity is made up of three parts representing, in turn, 43·7%, 28·7% and 27·6% of the total, the sum of these being exactly 100%. Rounded to the nearest whole number, the figures are 44%, 29% and 28%, which add up to 101%. It is good statistical practice to add a suitable footnote to the published figures when this possibility arises.

The limits of possible errors may be indicated in several different ways. The maximum error possible, the **tolerance**, on either side of the figure may be shown with a plus-or-minus sign ($\pm$). For example, if a figure given as 28·74 is known definitely to be in the range 28·72 to 28·76 it would be stated as $28·74 \pm 0·02$.

Alternatively, the actual limits of 28·72 and 28·76 may be stated, although the tolerance figure makes it easier to estimate at a glance the degree of accuracy. Either method indicates the **absolute error**: the actual magnitude of the greatest possible error rather than its size as a proportion of the whole.

Figures given on their own with no explicit statement of accuracy are assumed to be accurate to the last figure shown. So 28·74 standing alone is $28·74 \pm 0·005$; if the possible error in either direction were greater than 0·005, the figure 4 would be misleading and so ought to be omitted.

In the case of small observations, rounding to the nearest thousand or million produces problems. The population of a small country, for example, would appear as zero if the actual figure, before rounding to the nearest million, were less than half a million. To make clear the distinction between zero and non-zero observations, it is conventional to show a true zero by a dash (—) and numbers less than half the rounding unit as 0.

The implicit statement of accuracy noted above may be made explicit by indicating the last figure whose accuracy is guaranteed. The most common method is to state the total

number of guaranteed figures after (that is, to the right of) the
decimal point: so 28·74 is accurate to two decimal places or
2 dp. Alternatively, one may give the number of significant
figures. A significant figure is one which means what it says
—that is, it is not simply a figure included to show the magnitude of a number. Accordingly, 28·74 is correct to four significant figures (4 sig. fig. or 4 sf); so is 287 400 unless the
accuracy of one or both of the zeros can be guaranteed.
Correct to 4 sf, 287 400 could be as large as 287 449 or as
small as 287 350. Correct to 6 sf it could lie between
287 400·49 and 287 399·5. Conversely, 287 399·5 correct to
2 sf would be 290 000.

20 874 correct to 4 sf is 20 870: the zero between the 2 and
the 8 means what it says and is significant, while the other
zero does not—it is only there to show that the 7 is in the
'tens' position. Similarly, 0·028 74 is correct to 4 sf, as
the zero before the 2 indicates the order of magnitude in the
same way.

It is sometimes more useful to indicate the limits of error
as a proportion or percentage of the actual measurements.
This **relative error** is particularly useful when one wishes to
talk in general terms: for example, when describing the
accuracy of a slide rule as correct to 1 part in 1000, or 0·1 %.
The absolute error varies along the scale and cannot be
stated so easily.

The relative error may also be a more meaningful indicator. If a length of about 100 mm is said to be within ± 1
mm, the accuracy is not particularly impressive, but a length
of 1 km measured to the same absolute accuracy would be a
very different matter. The relative errors of 1 in 100 and 1 in
1 000 000 give a much clearer indication. The use of significant figures gives an approximate indication of relative
error—approximate only, because although 1·01 and 9·99
may both be correct to 3 sf, the relative error of the first is
almost ten times that of the second.

When expressing the accuracy of a rounded quantity in the form of a relative error, some care is necessary. For example, measurement of a quantity near 1000 to the nearest unit represents a proportional error of 1 in 2000, not 1 in 1000, since 'to the nearest unit' implies '± 0.5'. The maximum rounding error is, of course, a direct indication of the absolute accuracy of a measurement.

With very large and very small measurements, a way of writing numbers called **standard form** (sometimes called scientific notation) is convenient, both in itself and as a way of indicating accuracy. In this, a number such as 63 700 is shown as 6.37×10^4 (that is, $6.37 \times 10\,000$). The first part is a number between 1 and 10, which shows the number of significant figures without ambiguity, and the second part is a power of 10. Astronomers who use numbers like 2.997×10^8 clearly save space by writing them in this form, and to an experienced eye the magnitude of a number is easier to judge at a glance. Multiplication and division are easier than when using the conventional form; addition and subtraction are no more difficult. Numbers less than 1 take the form 7.193×10^{-6}: those unfamiliar with negative indices should read this to mean $7.193 \div 10^6$, or $0.000\,007\,193$.

So far it has been assumed that the original measurements have been objective ones, independent of human whim or memory. If this is not the case, errors of different kinds creep in as well: ages may be understated, numbers of cigarettes smoked may be rounded to the nearest five, items of income and expenditure may be left out. The problems which arise in this way are discussed in more detail in Chapters 7 and 9.

When measurements have to be added, multiplied, squared and so on, certain rules must be observed in order to keep track of the way the errors mount up.

First, consider addition. Clearly 5.76 ± 0.02 added to 3.21 ± 0.03 gives 8.97 ± 0.05—that is, the absolute errors are added. (If any of these examples give difficulty, consider the limits of the measurements: here, the two upper limits of 5·78 and 3·24 give 9·02, which is 8·97 + 0·05.) Relative errors are more difficult to deal with: $20 \pm 5\%$ added to $180 \pm 1\%$ gives a percentage error which depends on the 20 and the 180 as well as on the 5% and the 1%. It is best to work out the absolute errors of 1 and 1·8, giving a total error of 2·8, or $1\cdot4\%$ of 200.

For subtraction, analogy seems to suggest that the absolute errors should be subtracted, but this is not the case: as for addition, they are added. Thus subtracting $3\cdot21 \pm 0\cdot03$ from $5\cdot76 \pm 0\cdot02$ gives $2\cdot55 \pm 0\cdot05$. In order to give the upper limit of the answer, the *lower* limit of the number being subtracted is used with the upper limit of the other. A point to watch is that when a quantity is subtracted from another fairly similar in size, the *relative* error may increase enormously: if 246 000 is subtracted from 254 000, and each of these has a relative error of 1%, the difference of 8000 could be in error by as much as 5000, or over 60%.

When accuracy is stated by means of a number of decimal places or significant figures, the same principles apply but care is needed. The only really safe rule is to work out all the absolute errors and add them up. Thus 204·6 added to 31·278 would be shown as 235·9, as the second figure must be rounded to an absolute accuracy comparable with that of the first. Even now, the accuracy cannot be guaranteed as 4 sf because, if the 204·6 had been 204·56 before rounding, the total would have been nearer to 235·8 than 235·9. It is essential to avoid the mistake of giving the answer as 235·878, even though this would gain the coveted red tick in a junior school exercise book. To do so is to imply six-figure accuracy, when the correct total has 4 sf at most. Such 'spurious accuracy' is quite common in, for example, a population census when the total is sometimes left without any rounding at all, implying seven- or eight-figure accuracy.

When there are two or more figures which add to give an accurately known total, the errors are interrelated. A simple example is in opinion polls: if a sample of 1000 yields 520 in favour and 480 against, an error in one of these numbers must be balanced by a similar error in the other; the difference between the two is therefore wrong by twice as much as the original error.

When a long list of figures has to be added, the errors mount up. But, correspondingly, it becomes less and less likely that every one of the errors will take the pessimistic value which has been assumed for it, and that all of them will be biased in the same direction. It is this sort of situation which more advanced methods, requiring mathematical analysis to explain, can cover most effectively. They use the concept of probability: the basis for this is seen in the phrase 'less and less likely' above. The term 'standard error' indicates that such methods have been used; a similar term 'probable error' was once used in this context but is now obsolete.

Next we consider multiplication and division. By contrast with addition and subtraction, *relative* errors are easier to deal with. To see how the method works, consider two numbers, each with a given absolute error, say the length and breadth of a rectangle, to be multiplied in order to find the area. If the length is 400 ± 8 and the breadth is 300 ± 9, the maximum possible area is $408 \times 309 = 126\,072$, which is $120\,000 (= 400 \times 300)$ plus 6072. The lower limit is $114\,072$, or $120\,000 - 5928$. This seems to show little in the way of pattern, and even has different possible errors above and below the mean. But if the half-way point between the two errors is taken, which is 6000, an easy connection can be seen between the relative errors—2 % on the length and 3 % on the breadth add to give 5 % on the area. The two possible values of 6072 and 5928 actually give 5·06 % and 4·94 %, but these are near enough to 5 % for practical purposes.

The method may be explained by a diagram.

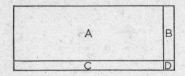

Suppose we increase the length of the rectangle by 2%. The initial area A is then increased by area B, which is 2% of A since it is a rectangle having the same breadth and 2% of the length. Now we increase the breadth by 3%, to include areas C and D. The area of C is 3% of A, so that B and C together add 5% to A. The area of D will be small compared with B or C, as the percentage errors themselves are normally small, and so it may be ignored. (Some readers may consider that actions of this kind are against the rules of the game, but in fact their judicious use is part of the art of statistics.)

The percentage error of the result is therefore found by adding the percentage errors of the quantities multiplied. If the errors are such as to reduce the length and breadth rather than increase them, the result is similar except that the total of areas B and C overstates the actual maximum error instead of understating it.

To give some idea of the effect of ignoring area D, suppose the initial errors were both 10%. The actual maximum errors in the product are then 21% for an overestimate and 19% for an underestimate, compared with 20% obtained by adding the initial errors. If both are 1%, the figures are 2·01% and 1·99% compared with 2%. In practice, many errors are of the order of 1%, particularly those involving calibrated instruments such as a pressure gauge or many types of electrical meters.

When figures are divided, the relative errors are once

again added. If this is not obvious by analogy with multiplication, a simple arithmetical example will demonstrate it. If three or more items are multiplied or divided, all the relative errors are added. Squaring counts as multiplying, so the relative error is doubled; for a square root, by reversing this process, the error is halved—one of the rare cases in which an answer is more accurate, in relative terms, than the data.

In advanced statistics, a margin of error is often specified rather like a tolerance, but with this difference—the limits are not absolute. Thus if an opinion survey of a random sample of 1000 voters shows 530 Nationalists and 470 Republicans, it does not follow that the proportion of Nationalists in the whole electorate is exactly 0·53, because of the uncertainties involved in sampling. Instead, the proportion would be given, using this advanced method, as '0·53 ± 0·03, with 95% confidence'. The confidence level is rather like a probability measure, although strictly speaking it is not one. For the present purpose it is sufficient to think of it in the same way as the phrase '99% certain' is used in everyday conversation. The method is mentioned again in Chapter 6.

This approach can also be used, on a much more subjective basis, when there is known to be a wide margin of error in an estimate, such as in predicting long-term changes on a basis of current values and trends. All too often—even by governments—a point estimate alone is given, with no indication of whether the error is of the order of 5% or 50%.

These methods all assume that numbers less than unity are shown in decimal form, as is usually the case. The accuracy of a fraction such as $\frac{3}{8}$ is difficult to show clearly, and faulty in principle since the use of a fraction implies exactness. Their use is less common now than it used to be.

Finally, it is worth stressing as a point of principle a matter which was mentioned earlier—that no quantity measured on a continuous scale can ever be found with total accuracy, and that this covers every kind of physical measurement except those made by counting. Thus it is impossible to carry out an instruction such as 'Cut a length of precisely 10 cm from this piece of string'. The principle has been well understood for a long time:

> Shed thou no blood; nor cut thou less nor more
> But just a pound of flesh: if thou tak'st more
> Or less than a just pound—be it but so much
> As makes it light or heavy in the substance,
> Or the division of the twentieth part
> Of one poor scruple: nay, if the scale do turn
> But in the estimation of a hair—
> Thou diest, and all thy goods are confiscate.
>   (Shakespeare, *The Merchant of Venice*, Act IV.)

For a more up-to-date example, a kilogramme packet of sugar cannot contain exactly 1 kg and, if the manufacturer wishes to be sure for legal reasons that no packet contains less than 1 kg, he must set his packing machine at a nominal level above 1 kg, by an amount equal to the tolerance at which the machine operates. Shylock's problem was that the laws of Venice would not allow him to go either above or below the amount stated in the bond.

An allied point is that a measurement can never be stated to be zero, again unless achieved by counting. Thus a legal requirement that a supply of drinking water should be entirely free of contamination by lead would be impossible to observe—unless a testing method could be developed which detected individual molecules, which is not within the range of practical possibility. In technical language, to bring both sides of the subject together, 'zero tolerances' are un-workable, and in every case the practical limit is based on the sensitivity of the available testing methods.

4 Changing many figures into few

No statistical analysis can draw out from a set of figures more information than it contains, or any greater accuracy than the figures possess. It can, of course, show the information and its accuracy in a more comprehensible form, as we shall see.

But the first task of the statistician is to decide what records to keep—the measurements to be made, and the optimum accuracy for each. At least, it is a job which ought to be within the statistician's control, though often it is not. So, useful measurements are sometimes left out, or useless ones left in; readings may not be taken accurately enough, or money may be wasted in achieving a high degree of accuracy to no purpose.

When the choice of measurements has been made, the next and obvious aim is to devise a foolproof system for recording the figures. These figures form what is known as the raw data for the subsequent analysis, and in one sense this first stage in the process is the most important: if a mistake is suspected in a calculation later on, it is always possible to check, but the conditions in which the original measurement was made may be impossible to reproduce later. The recording of the raw data is all too often left to a junior member of an office staff, or if done by a senior one may be regarded as less important than his other responsibilities. Less urgent it may indeed be, but mistakes made at this stage may have serious consequences later on.

The first stage in the collation and analysis of raw data is to set out the information in a table. At this first stage the tabulation will preserve the standard of accuracy of the data. The table provides a convenient reference source for those who wish to check the original analysis for themselves or for those who wish to use the data to carry out new analyses.

For the purpose of analysis the figures are usually grouped in whatever happens to be the most convenient way. Suppose we have the ages of 10 000 people in a community, and we wish to form a general impression of the balance of numbers in the various age groups—how many of school age, how many over-sixties and so on. To give over 90 figures, one for each year, would be more confusing than otherwise, and so they are simplified into groups or **classes**. One method would be to combine data in 10-year groups: 0–9, 10–19 and so on. (This is slightly better than 0–10, 11–20, etc. since in counting up it is then only necessary to ignore the last digit in order to place each item in its correct group.) The range covered by each group (here 10) is the **class interval**, and the number of observations in a particular group is the **frequency**.

In both methods care is necessary with the end groups. If ages are measured to the nearest year, the lowest group covers $9\frac{1}{2}$ years if 0–9 and $10\frac{1}{2}$ years if 0–10; if they are measured in completed years, the lowest group covers 10 or 11 years respectively. The top group has to be left open-ended, either 80+ or 90+, and spans an uncertain number of years.

It would probably be better in this case to use a smaller class interval, 5 years instead of 10. The choice of interval depends on the amount and quality of the data, and the use to which the answer is to be put, but as a general rule the number of groups should be between 6 and 20. Sometimes a case can be made out for using different class intervals together—narrow when observations are tightly packed, wider when the density of the observations is low.

The intervals should be chosen in such a way that no actual observation can coincide with the dividing points. Thus if the observations are rounded to 1 dp, the class boundaries would be chosen as, say, 9·95, 19·95, 29·95, etc., rather than 10, 20, 30. Then there is no doubt about which class an observation is in.

As with more complex calculations, some of the information conveyed by the raw data is lost in the grouping process: we know, say, the number of children in the 5–9 group, but the number of 6-year-olds is not to be recorded. For some purposes, the loss might be important (and so the raw or tabulated data should never be thrown away), but for many others it is more than balanced by the increased clarity of the data as a whole. If further calculation is to follow, perhaps to find a mean age for the group, it can usually be arranged so that most (if not all) of the errors are compensating rather than biased, thus keeping the total relative error small. The calculations themselves are much easier.

It may be mentioned here that the habit of using a comma to divide large numbers into groups of three digits, working backwards from the decimal point, is no longer recommended because the comma is used in some countries to mark the decimal point itself. Instead, a thin space is used to mark each three digits before and after the decimal point. This was recommended by an international conference in 1948, and published as a British Standard in 1953; even so, many official publications still use the comma. The convention adopted in this book (as allowed by the Standard) is to omit the space in four-digit numbers and when there are four decimal places, except in tabular work.

Proportions and percentages

The basic ideas of proportion and percentage will be familiar to most readers. The proportion, essentially, is a fraction: if a firm has 200 employees of whom 120 are men, then the proportion of men is $\frac{120}{200}$ or $\frac{6}{10}$. Proportions are normally used for comparing the size of a particular group with that of a larger group of which it forms a part—here, the male employees with all employees. Its value when used in this way is less than 1.

A percentage is, in principle, a proportion scaled up by 100. In the example above, the percentage of men is $\frac{6}{10} \times 100 = 60\%$. Percentages are also used as a measure of change, and in comparisons generally: a figure of 480, compared with one of 400 from another time or another place, shows a change of 80 in 400, or 20 in every 100, which becomes 20%. A percentage change, if it is an increase, can therefore be more than 100%; but see below for a note concerning very large changes.

Comparisons are sometimes also expressed by means of a ratio, particularly when two separate groups are being compared. The numbers of men and women employees in the example above are in the ratio 120 to 80, or 3 to 2, and this may be written 3:2. The form $n:1$ is often convenient; here the ratio 1·5:1 shows that there are one and a half male employees for every female one. Obviously, the figures may be looked at from the other point of view also, and the ratio $1:\frac{2}{3}$ would show that the number of female employees was two-thirds the number of the male ones.

Percentages should not be used to express changes in quantities which are themselves percentages. Thus if a profit expressed as a percentage of working capital goes up from 12% to 15%, some way other than a percentage must be used to describe the increase, since figures of 3% and 25% could both be quoted with some justification, and the result would be confusion.

Also, percentages are not the best method in the case of very large increases. If a work of art appreciates from £2000 to £40 000, a better impression of the magnitude of the increase is given by saying that the new value is twenty times the old, rather than that its value has increased by 1900%. (This figure would often be incorrectly quoted as 2000% anyhow.)

Percentages should not be compared with each other unless they are related either to the same quantity or to directly comparable ones. Thus a company which increases its output by 25% compared with the previous year is not necessarily doing better than one with a 5% increase: it may have had an unusually low output in the previous year because of a strike, or it may be a small company just entering the market. Likewise, percentages should not be added or averaged unless they are related to the same reference figure.

Percentage comparisons with earlier figures may be made to look more or less favourable by varying the choice of first date. Thus the directors of a company whose profits in successive years were £60 000, £50 000 and £57 000 might call the latest figure a 14% rise over last year's; less charitable critics would call it a 5% decrease in two years.

The use of percentages which are not derived from objective measurements is, at best, unhelpful. To say that intelligence is 80% derived from heredity and 20% from environment (or the other way round) is to beg so many questions as to be positively dangerous.

But in certain cases a proportion or percentage is the only correct basis for a comparison. It may be true that more people are killed each year when crossing the road than while climbing mountains, but this does not mean that climbing is the less dangerous activity of the two. When related to the numbers taking part in each, the *proportion* killed while crossing the road is far smaller, and this gives a more accurate indication of the relative degrees of risk.

Using two percentages in succession can lead to apparent inconsistencies. If a quantity increases by 10% and then falls again by 10%, it does not return to its original size. For example, 200 goes up to 220; 10% of this is 22, so the final figure is 198. Likewise, two 10% increases are equivalent to 21% overall, not 20%. (The analogy with relative errors in multiplication will be noticed—see Chapter 3.) The magnitude of the inconsistencies increases as the percentages themselves increase: thus two 40% increases make a 96% increase, not 80%, and two 100% increases make 300% not 200%. The same is true for more than two successive increases, as when a population growing at 2% per annum doubles its size in 35 years, not 50. Two successive decreases are less in total than the added percentages: thus a 50% drop followed by another of the same size gives a 75% drop in all

A proportion or percentage should never be accepted at face value unless the reference quantity is stated. An advertisement claiming that 9 out of 10 cats prefer a particular brand of pet food *may* mean what it appears to mean, but only if a statistically satisfactory sample of around 1000 cats or more has been tested under controlled conditions. In the absence of positive evidence, the advertiser's claim must be rated as highly suspect. The vagueness of the statement probably conceals a vagueness in method also.

For basically the same reasons, a proportion derived from one set of measurements should only be taken to apply also in a different situation by someone who knows what he is doing. If 60% of the adult population of a town are shown to be in favour of capital punishment, it does not follow directly that 60% of the whole country, or 60% of the population of some other town, are also in favour. There may be some differences, because of the different characteristics of the two areas being compared; and there may be large differences as well, perhaps as a result of a recent murder case in the town. Equally, there may not, but the onus of proof is on the statistician, not on his readers. The techniques required to provide such proofs are discussed in Chapter 7.

Representative values

In competitive rowing it is considered an advantage to a crew if the oarsmen are heavier than their opponents and, in order to express a comparison between two crews, it is helpful to be able to use a single measure for each crew's weight.

The easiest way is to work out the two totals, but this has the disadvantage that the size of each total does not convey an immediate impression of being either heavy or light, because it is not related to the reader's knowledge of people's weights. The reason is, of course, that the size of the total depends on the number in the crew as well as on their weights. Instead, imagine an eight-man crew whose total weight is equal to that of the actual crew, and in which all eight men have the same weight. Then take this weight as a representative measure for the weight of the actual crew. The figure found in this way is the **average** or **mean** (strictly speaking, the arithmetic mean; there are other kinds).

This explanation indicates the method of calculation. If there are eight imaginary men of equal weight forming a crew of the same total weight as the actual crew, the weight of each one of them will be one-eighth of the total. Hence, to find an average, add all the measurements and divide by their number.

The use of the average is covered in more detail later on but, before this, some other and less important types of representative measure must be mentioned.

One could convey an immediate impression of the heaviness or otherwise of the rowing eight by setting the men in order of weight, with the heaviest at one end of the line and the lightest at the other, and taking the weight of the man half-way along the line. If there were nine men, the fifth in order would be the middle one, with four on each side of him; with eight it is necessary to take the average of the weights of the fourth and fifth.

The measure thus found is the **median**—roughly speaking, the middle item in size order. It is easy to calculate, but for most purposes is not such a good representative measure as the average. If a rowing crew included five fairly light members, the heaviest two of these would determine the median value irrespective of whether the other three were only a little heavier or very much so.

There is one case, however, in which the median is a better measure: when the measurements are much more widely spaced at one end of the range of values than the other. In the salary structure of a large firm, the cash difference between the salaries of those on the lowest pay scale and those roughly half-way up the promotion ladder is usually much less than the difference between the half-way group and the managing director. So the salaries at the top end of the scale make a disproportionately large contribution to the size of the average. The median gives a better impression of the salary of an individual who is half-way to the top.

To find which item is the median, add one to the number of measurements and divide by two, so that for 47 items the median is the 24th in order. If there is an odd half, take the average of the two each side. When the measurements are grouped by size, the median is found by proportion within the group which contains it: if the median measurement is the 62nd in order, and measurements 54 to 79 are in, say, the 50–59 kg group of mass to the nearest kilogramme, the median is $49\frac{1}{2}$ plus $\frac{9}{26}$ of 10, or about 53 kg. The $49\frac{1}{2}$ is the lower limit of the 50–59 group before rounding, and the 62nd item is ninth of the 26 in the group. The median cannot be used as a precise measure, so that high accuracy is pointless—even taking the $49\frac{1}{2}$ inaccurately as 50 would be sufficient in practice.

When the number of measurements is large, and when they show a clear pattern with many observations in one part of the range (usually around the middle), a third type of representative measure may be used. It is no use when there are only a few measurements, as in the rowing crew—50 is about the minimum. Suppose, for example, that two dice were thrown together, about 100 times, and the total score was recorded each time. The double-one and double-six scores would be less common than totals in the 5–9 range which result from several combinations of individual scores. One total—probably the 7—would probably occur more often than any other, and this particular value is the **mode**. If the measurements are grouped, with a uniform class interval, the **modal class** is the one containing most observations. In a large set of grouped observations, it is possible to make an estimate of the mode itself from the numbers in the three biggest groups, but it is rarely worth the trouble. The mode, in fact, is not usually a good choice of representative measure: it is far too dependent on chance irregularities in the pattern of measurements. Neither the median nor, still less, the average is subject to such variations.

The mean

The terms 'average' and 'mean' are interchangeable, and their uses in composite terms such as 'moving average' and 'weighted mean' indicate a difference only by convention and not in fact. 'Average' is generally more common, but 'mean' is the word used more often by statisticians. On its own it implies the arithmetic mean, as defined in the last section; the geometric or harmonic mean is specified by its full name when required.

In calculating a mean, it is sometimes possible to use methods which shorten the work, rather than add all the measurements and then divide by the number of them, as the definition requires. Details of these methods will be found in books with a more mathematical approach, and are worth learning by readers who frequently have to do such calculations.

When measurements have been grouped, it has to be assumed that all those in one particular group are at the centre-point of the group—for example, if heights of individuals are grouped with a class interval of 10 cm, those in the

160–69 group when measured to the nearest centimetre are all assumed to have height 164·5 cm. The errors in this process are unbiased, so that the considerable saving of labour when dealing with a long list of measurements is offset by only a very small error in the mean. A slight difficulty may arise if one group is open-ended, as in a top age group; the best way out is to fix an arbitrary limit on a common-sense basis. Since the number of observations in such a group is usually small, the resulting error will be small also.

For almost all purposes, the mean is the most truly representative single figure for a given set of measurements. There are times when a single figure is not really sufficient, as will be mentioned later, but even then the mean will still be one of the two or three measures used. Its value is affected by the size of every one of the original measurements, and this is not true of the median or mode. The mean is fundamental to most calculations in advanced statistics.

The mean of two or more means cannot be calculated directly, unless the number of observations in each of the groups is the same. For example, if the 24 boys in a class have a mean height of 159 cm and the 8 girls have a mean height of 151 cm, the mean height of the whole class is not 155 cm. Instead, the total of the boys' heights must be found as 3816 cm, and that of the girls as 1208 cm; the sum of these is then divided by 32 to give a mean of 157 cm. If there had been 16 each of boys and girls, the simpler method of adding 159 to 151 and dividing by 2 would have given the correct answer, which in this case is 155 cm.

Likewise, an average of two rates cannot be found directly. A number of apples at 16p per kg mixed with others at 20p per kg is not the equivalent of 18p per kg overall, unless the weights of the two types are the same. The best-known catch question along these lines is to ask the average speed of a swimmer who goes 3 km upstream at 2 km/h and back at 6 km/h—the answer being that he travels 6 km in a total time of 2 hours, at an average therefore of 3 km/h. The mean of 2 and 6 gives the wrong answer. The correct mean here is the harmonic mean, but this is rarely used in statistical work.

The geometric mean

When a quantity such as a human population grows under uniform conditions, at so many per cent per year, the measured increase itself increases each year, as it is a fixed percentage of an increasing quantity. If a population doubles, the measured increase per year doubles also.

Suppose the population of a country was 4 million in 1800 and 9 million in 1900. Then the figure for 1850, assuming stable conditions, would not have been the arithmetic mean of $6\frac{1}{2}$ million since this makes the measured increase in each 50-year period the same, $2\frac{1}{2}$ million. The correct value is 6 million, which makes the *percentage* increases the same, as required: 50% from 4 million to 6, and the same again from 6 to 9. This kind of mean is a **geometric mean**. The method of calculation is to multiply the two numbers whose mean is to be found, and to take the square root of the result. The geometric mean is sometimes used as an alternative to the arithmetic mean. For more than two items the method is to multiply all *n* of them together and take the *n*th root of the result. The geometric mean is always less than the arithmetic mean, and the principal effect of using it rather than the other is that the distortion produced by a single exceptional measurement is usually rather smaller. But it cannot be used if any of the measurements is zero or negative, and a single very small measurement may still produce a distorted result.

Moving averages

The arithmetic mean may be adapted to suit certain special purposes. For example, the moving average can be used to show a trend in a series of measurements over a period of time, such as monthly unemployment or quarterly output. In such statistics there is often a seasonal trend—some forms of employment, for instance, cover the summer months only, and so it is normal for the summer unemployment figures to be lower. What is usually of most interest, therefore, is any possible non-seasonal trend, and the moving average is a way of isolating it.

If the measurement is to be taken at quarterly intervals, the average of the first four quarters is found to start with. Then the average is found of the second to fifth quarters— that is, the first quarter is left out and replaced by the fifth. After this the process is repeated—third to sixth, fourth to seventh and so on. It will be seen that the effect each time is to set the latest available figure in place of that for the corresponding period one year before, and so the change in the average will be determined by the difference between these two figures.

To make this clearer, suppose the first five quarters showed outputs of 50, 70, 60, 40 and 58 units. The ups and downs of the first four quarters are of no particular concern for the present purpose, and what is more important is that the output for the fifth quarter is better than that of the first, which was the corresponding period one year before. The average of the first four quarters is 55, and the average of the second to the fifth is 57. (It will be seen that the difference is one-quarter of the increase from 50 to 58; this can be used as a short cut in calculation.) Thus the moving average gives a continuous comparison with the last period for which a true comparison can be made.

For a monthly moving average, twelve observations are taken together, and similarly for other periods of time. To some extent the process allows for the differing lengths of months, but the figures may be adjusted to allow properly for this. A moving average over a longer term—say, ten years in steps of one year—may be used to show the long-term trend when there are wide variations up and down in the figures for the separate years.

Weighted means

Consider an examination in two parts, Theory and Practical, each marked out of 100. If the overall result is worked out simply by adding the two marks together, a candidate with 30% on the theory and 60% on the practical would of course gain the same total as another with 60% and 30% respectively. But suppose the examiner decided that the theory paper was more important, and should count twice as much as the other. The marks for this paper are therefore doubled, so that the first candidate gets 120 marks altogether out of the new total of 300, and the second 150. The second candidate thus gains more credit for his high marks in the theory paper, as expected.

The overall percentage marks of the two candidates are 40% and 50% respectively, and each of these figures is a kind of average of the marks on the separate papers—but an average which gives more 'weight' or importance to one figure rather than the other. In this case the 'weights' are 2 for the theory paper and 1 for the practical. An average calculated in this way is known as a weighted mean.

The method of calculation, in general terms, is as follows. Multiply each quantity by its 'weight'; then add; finally divide, not by the number of observations, as for the arithmetic mean, but by the total of the weights. For the first candidate in this example, twice 30 is added to 60, and the result divided by 3 to give 40%, as before.

Sometimes the weights are derived directly from the number of observations. Consider the example used earlier in which there were 24 boys of mean height 159 cm and 8 girls of mean height 151 cm. The weights used are 24 and 8 respectively, and the mean height of the whole group works out as 157 cm as before.

A simple mechanical analogy may help in understanding the process, when the weighted mean of two quantities (rather than more) is to be found. Imagine a metre rule (assumed weightless) with 100 g masses hung at the 30 and 60 cm points to represent the marks in the two exam papers. Then the arithmetic mean is at the point of balance between the two, that is 45 cm or 45%. For a weighted mean giving twice the importance to the theory paper, the mark for this paper is represented by *two* 100 g masses. The point of balance is now twice as far from the 60 cm point as from the 30, by the laws of mechanics. The weighted mean is thus 40%, since 40 cm is the point of balance, one-third of the way between 30 and 60 cm. The reader may like to work out the mean height of the 32 children by this method.

Index numbers

Perhaps the best-known of all Government statistics is the Index of Retail Prices, commonly (though inaccurately) called the cost-of-living index. This is only one of a number of indices, covering subjects such as production levels, sales figures and commodity prices, which are published at monthly, quarterly or annual intervals. Stock market indices are similar in principle. The basic idea of an index of this kind is that of the weighted mean. But the problems of construction and interpretation can in practice become very complex, and this sets a limit to their usefulness to the non-specialist. Index numbers have the quality of appearing to be very simple—and indeed they are, in the sense that what they say can easily be understood. But whether what they say is an accurate reflection of the true state of affairs is another matter.

The Retail Price Index is designed to reflect changes in

the prices of goods and services on which the average family spends its income. Expenditure is divided into sections such as food, housing, transport, clothing, fuel, alcohol, tobacco and consumer durables, with a suitable weight allotted to each for the calculation of a weighted mean. Thus if food or fuel prices go up (or, rarely, down), a small change in the index follows. Since food is a bigger item than fuel in a family's total expenses, food prices are weighted more heavily than fuel, and so on. The values of the index are published at monthly intervals.

The problems stem from two sources. First, the average family alters its pattern of expenditure, over a fairly long period, for what are basically social reasons: fashions in clothes change, frozen foods take the place of some fresh ones, coal fires give way to central heating. This means that the weightings have to be revised, either every year or at longer intervals, which limits the accuracy of comparisons between different times. Changes in the true cost of living over a period of, say, ten years or more are difficult to measure by any method, other than in fairly general terms. Over periods of the order of a century or more, only the prices of items like beer and butter, which have continued without much alteration, give any idea at all of how things have changed.

Secondly, shorter-term variations in the pattern of expenditure happen as a result of the very changes that the index is trying to record—and this is where the trouble really starts. In order to show what happens, let us invent an index to be called the Cost of Bread-spreading Index, which is designed to show the price movements in butter and margarine.

The standard of comparison is the pattern of prices at a particular time, the beginning of a year which we will call year 1, or the base year. Suppose the prices of butter and margarine, in pence per kilogramme, are then 56 and 42 respectively; suppose also that the necessary survey shows that in these conditions the average family spends 40% of its butter/margarine budget on butter and 60% on margarine. Then the weightings will be taken as 40 and 60 respectively. (Or 400 and 600; the ratio is important, not the actual numbers, and it is conventional to make the sum of the weightings either 100 or 1000.) In year 1 the index is given the nominal value of 100.

Then in year 2 let us introduce a shortage of dairy products, so that the price of butter goes up to 65. Margarine goes down slightly to 40. The new index, using weightings 40 and 60, is calculated as follows:

$$\frac{65 \times 40 + 40 \times 60}{56 \times 40 + 42 \times 60} \times 100 = 105 \cdot 0.$$

Thus the value of the index for year 2 is 5% above that of year 1.

Readers who choose to skip over the details of the calculations will still be able to follow the general line of argument but, for those who want to think it through, the method goes like this. The initial prices are 56 and 42. The simple average of these is $(56 + 42) \div 2$, or 49; but instead of this we require a weighted mean which allows more weight to the margarine price, so that the result will be less than 49. This mean, worked out as explained above, is equal to $(56 \times 40 + 42 \times 60) \div 100$, or 47·6. Similarly, that for year 2 is $(65 \times 40 + 40 \times 60) \div 100$, which equals 50·0. The ratio between these two is 1 : 1·050, or 100 : 105·0.

The result is not difficult to understand, and it seems to imply that the average family is now having to spend about 5% more on its bread-spreading than last year. But this is not so. Consider a family which in year 1 bought 1¼ kg of butter and 2½ kg of margarine each month. This would have cost 70p for butter and £1·05 for margarine, a total of £1·75;

the ratio of these is 40:60, in agreement with the weightings. When butter goes up by 9p and margarine down by 2p, the family will buy less butter—suppose the new amount is $\frac{1}{2}$ kg, and that they buy $3\frac{1}{4}$ kg of margarine, thus keeping the total quantity constant at $3\frac{3}{4}$ kg. This costs $32\frac{1}{2}$p plus 130p, giving a total of £1·62$\frac{1}{2}$ which is *less* than before, not 5% more. Of course, these figures are exaggerated somewhat—but a family varies its expenditure pattern as a result of changes in income as well as changes in prices, and a price index is not designed to take such matters into account. Discrepancies of this kind do therefore occur, even if they are not in practice as glaring as this.

It is not difficult to alter the method of construction of the index so as to take into account the changes in the expenditure pattern: one simply has to use different sets of weightings for the two years. But such an index would be valueless in assessing the true changes in the cost of living—by implication, a particular standard of living, not one which is progressively reduced by price increases. It will be seen that the interrelationships between the different factors involved make the situation a very complex one, and to try to reduce the changes to the values of a single numerical quantity is to attempt something which is difficult if not impossible.

It is possible, however, to make one change which goes some way towards taking expenditure into account. This is to use the pattern of expenditure in year 2 to fix the weightings, rather than that of year 1. Such a procedure is only possible if information on expenditure is obtainable more or less instantaneously, and in practice this takes some time to prepare. Separate names are used to distinguish between the two methods: a Laspeyres index is one which uses the old quantities to calculate the weights, and a Paasche index is one which uses the new. The two may be combined, by taking their geometric mean, to give a third type which is known as Fisher's Ideal Index after its inventor Irving

Fisher; this has certain technical merits which go some way at least towards justifying its name.

In the example used here, the family's expenditure on butter and margarine in year 2 is in the ratio $32\frac{1}{2}:130$, or $1:4$. The weightings are therefore 20 and 80, compared with 40 and 60 for the Laspeyres index. The value of the Paasche index for year 2 is therefore

$$\frac{65 \times 20 + 40 \times 80}{56 \times 20 + 42 \times 80} \times 100 = 100 \cdot 4.$$

As expected, this is closer to 100 than the Laspeyres index value of $105 \cdot 0$.

The next stage is to work out what happens in year 3. Suppose now that butter goes up slightly to 68, and that margarine rises sharply to 51. The family now buys, say, $\frac{3}{4}$ kg of butter and 3 kg of margarine, costing $51 + 153\text{p} = £2 \cdot 04$; the weighting is $51:153$, which equals $25:75$. Related to year 1, and again with the weights based on the current year, the Paasche index for year 3 is

$$\frac{68 \times 25 + 51 \times 75}{56 \times 25 + 42 \times 75} \times 100 = 121 \cdot 4.$$

This compares with an actual increase in expenditure of $16 \cdot 6\%$, which as before is less than the rise in the index because the family has adjusted its expenditure pattern, as a reaction to the changing balance between the prices.

The problem is now to relate year 3 to year 2. A direct comparison between the two values of the Paasche index is impossible, because the values have been based on different weightings. (This would not have happened if a Laspeyres index had been used, but instead there would have been some distortion of the true picture because the weightings in year 3 would have been two years old.) So a new index is calculated by relating year 3 directly to year 2. This has the value

$$\frac{68 \times 25 + 51 \times 75}{65 \times 25 + 40 \times 75} \times 100 = 119 \cdot 5.$$

So far so good, and it only remains to relate year 3 to year 1. This is done by a method analagous to that for finding the result of successive percentage changes: the value just found counts as a 19·5% increase, so that the value of 100·4 for year 2 is multiplied by 119·5/100, giving 120·0. When the weights are revised each year in this way, the result is called a chain index, and so the figure just found is the value of a chain Paasche index for year 3, using year 1 as the base year.

When a chain method (either Paasche or Laspeyres) is employed, the annual revision to the weightings enables changes to be made gradually, whether the reasons for the changes are social or economic. This avoids the awkward discontinuity caused by a major revision at longer intervals. But revision in this way also makes comparisons over a long period unreliable. To give a specific if somewhat unlikely example: if the prices and weightings should happen to return after a few years to their values at the base year, there is no reason why the index should return to 100. If this does happen in year 4 to the index used here, the value comes in fact to 98·8. The difference may seem small, but the changes in the Retail Price Index are increasingly being used as a basis for wage negotiations, not just in general terms but by direct relationship. In such conditions a one-point rise can be important.

In the case of the Retail Price Index, the method used is that of the chain Laspeyres index. Since this is the type which is rather less affected by changes in the expenditure pattern than the Paasche, it is more truly a measure of the change in the cost of living at a particular standard. Most of the weights are derived from the results of the Family Expenditure

Survey described in Chapter 2; for large items such as cars and furniture a separate survey is taken, in which short-term changes are smoothed using a three-year moving average. The reason why the index should not, strictly speaking, be described as a cost-of-living index is that it leaves out of account all direct taxation and some other items such as insurance and mortgages.

When related to the base year, an index shows the rise or fall directly as a percentage. For comparisons other than this, the change may be expressed either directly as a number of 'points' or by calculation as a percentage. For changes when the index is over 100, the percentage is the smaller of the two.

Comparisons between indices for different quantities or regions are inconvenient if the base years are different. There is some international standardisation on this, and all the indices for the United Kingdom now have 1970 as the base year.

Pointing out the defects of an index number is all too easy, and is not being done here in order to criticise those who construct and use them. In spite of the disadvantages, there is no substitute which is so readily comprehensible and, provided that the changes being measured are both reasonably small and fairly short-term, indices serve a useful purpose and will no doubt continue to do so. The simple fact that they are both comprehensive in coverage and easy to understand is a merit which makes their defects quite acceptable.

Measures of dispersion

Consider the problem of a manufacturer who makes raincoats in various sizes to suit men of different heights—150 to 160 cm, 160 to 170 cm and so on. He receives an export order for 20 000 raincoats, to be made in appropriate numbers for each size. What information will he require about the heights of the men in the country concerned?

Clearly the mean alone is not sufficient. A mean of, say, 175 cm is no help in deciding how many there should be in the 160–170 cm size. What is required is a measure of the dispersion of the heights—that is, a way of showing whether they are clustered narrowly round the mean or ranged more widely. Just as there are various forms of representative measure such as the mean and median, so also there are various ways of measuring dispersion, of which one in particular is much the most useful.

First and simplest is the **range**. This is the difference between the largest and smallest measurements. It is very simple to calculate, and gives a useful first impression of the magnitude of the dispersion. But for detailed analysis its value is small, since it takes account of only two measurements. The raincoat manufacturer would require more information than this.

A measure similar to the median can also be used. The median is the observation half-way down the list when they are placed in order of size, and the measure used for the dispersion is worked out from the observations one-quarter and three-quarters of the way down the list. These points are called the **quartiles**, lower and upper. If the measurements are clustered around the mean, the quartiles will be close together, and in a wider spread of measurements they will be further apart. The actual measure used is half the difference between the two, called the semi-interquartile range or quartile deviation. These measures are not as easy to interpret as the median itself, and often do not give sufficient information anyhow; for the raincoat manufacturer they would be an improvement on the range, but still not good enough. One reason is that the quartiles are independent of the actual values of the highest and lowest quarters of the ordered list of measurements. (See also Figure 9, page 65.)

Percentiles may be defined in a similar way, as the points dividing the range into 100 equal parts. The 25th and 75th percentiles correspond with the lower and upper quartiles, and the 50th with the median. Measures derived from percentiles—particularly the 10–90 percentile range—are common in sociological work. Just as for the quartiles, the main disadvantage is that the extreme values are not taken into account.

The other and far superior measure, the standard deviation, is not so easy to explain. Like that of the mean, the method of calculation takes into account the size of every measurement, though this is not the only reason why it is the best measure of dispersion. It marks very precisely the point at which the kind of statistics which this book is designed to cover comes to an end and advanced work begins. It therefore finds little mention in the following chapters, and even the brief explanation which follows may be omitted if desired.

One way of ensuring that every measurement is taken into account is as follows. If the dispersion is low, with the observations clustered closely around the mean, the average difference between the observations and the mean is low also; with greater dispersion, the figure is higher. The differences, found by subtracting the mean from each observation in turn, are **deviations**; measurements below the mean have negative deviations. To find the average deviation, the negative signs are omitted and the figures averaged in the usual way. The mean deviation is one possible measure of dispersion.

The need to omit negative signs gives rise to certain practical difficulties and, for this and other reasons, the mean deviation is rarely used. But the calculation of the standard deviation follows the same general lines. It also is a mean deviation from the mean, but calculated in a rather special way.

Suppose you were asked to find the average size of three carpets, 1·5 m square, 2·1 m square and 3·9 m square respectively. One way would be to add the three figures together and divide by 3, giving 2·5 m. But clearly this would not be very sensible, as the quantity which should be averaged is the area, represented by the *square* of the length of the side. The steps in the calculation are :

  (i) Square each figure, giving 2·25, 4·41 and 15·21.
  (ii) Add, giving 21·87.
  (iii) Divide by the number of measurements (3), giving 7·29.
  (iv) Take the square root of the result, giving 2·7.

The average size is therefore that of a carpet 2·7 m square, not 2·5 m square.

This is exactly the process used in the calculation of a standard deviation, the lengths of the sides being replaced by the deviations from the mean. In mathematical language, the standard deviation is the square root of the mean squared deviation from the mean or, more briefly, the root mean square deviation. It will be seen that the squaring process removes the problem of the negative signs.

A very simple example of the calculation of a standard deviation is as follows. If the observed quantities are 15, 17, 19, 19, 21, 21, 23 and 25, their sum is 160 and their mean 20. The deviations are respectively −5, −3, −1, −1, 1, 1, 3 and 5. The squared deviations are 25, 9, 1, 1, 1, 1, 9 and 25, whose sum is 72 and mean 9. The square root of 9 is 3, and this is the standard deviation.

Whether the standard deviation is a sufficient measure of the dispersion in a particular case depends on what else is known about the pattern of the measurements. Quite often there is a clear pattern, with a high proportion clustered round the mean and fewer towards the tails—for instance, the raincoat manufacturer would certainly make more of the medium size than of the very small or very large sizes. In certain situations (discussed briefly in Chapter 6) it is possible to describe the pattern mathematically, so that all the percentiles can be calculated from the values of the mean and the standard deviation alone. And many others give a pattern which is a fair approximation to this, the heights of adult men being one example. When this does happen, the points which are one standard deviation below and above the mean are at approximately the 16th and 84th percentiles —that is, they will include between them roughly the central two-thirds of all the measurements. About 95% of the measurements will similarly, under these conditions, be within twice the standard deviation of the mean. Figure 20, on page 82, illustrates this pattern.

It is often useful to refer to a particular observation in terms of its position in the pattern of measurements. We often do this in a qualitative manner by saying of someone that he is of about average height, or that he is unusually tall. Referring again to Figure 20, it will be seen that this can be done by stating the number of standard deviations above or below the mean. The deviation of the required measurement is divided by the standard deviation, and the result is a **standard measurement**. An observation with a small standard measurement is described colloquially as 'about average', while a standard measurement over 3 is exceptional. Observations less than the mean have negative standard measurements.

This, alas, is as far as one can go in an introductory book. The reader who wishes to deepen his understanding of what statistics is all about should find much of interest in the following pages, but at several points there will appear a barrier to further progress in the form of the mathematics of the standard deviation. Those who go on to more advanced books, and are able to remove the obstruction, will find much more of interest beyond.

5 Figures with an impact

Anyone attempting to include in his library all the statistical information published each year would soon find himself running out of space—there is simply too much of it. Even a small part would still be too much for any one person to understand and assimilate, and so some degree of selection and simplification is essential, using the methods described in the last chapter.

But there is another reason why simplification is necessary. The development of rapid communication of news of the day by television has led in turn to new ways of getting information across, not only to the viewer and listener but to the reader as well. Television is not in fact a particularly suitable medium for conveying statistical information—the plight of one homeless family can be shown more vividly than any statistics, but solid information about the *extent* of the problem is just as important. It is possible that the visual impact of news film of wars and disasters has made it more difficult than before for mere statistics to make their own impact. But what is quite certain is that the ordinary man has become accustomed to instant communication and that, as a result, those who wish to get statistical information across to him must use simple and direct methods. Often these methods will make use of graphs and pictures, which are the subject of this chapter.

Simplification always leads to some degree of distortion—great or small, calculated or unavoidable. In advertising, particularly, those who draw the graphs are trying to present the information in such a way as to show their product in the best light, and the reader must be on his guard.

Time graphs

By far the commonest method of visual presentation is a graph showing how a quantity varies over a period of time. Two principal versions are possible. When the quantity is a continuously varying total, such as unemployment, and is measured on selected and evenly spaced dates, a series of points is joined up by straight lines, as in Figure 1. If the

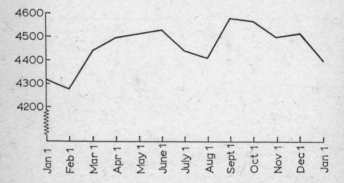

Figure 1   Time series of the values of a continuously changing quantity

points are very close together, a curve may be used instead. Alternatively, when the quantity is measured as a series of separate totals for successive equal periods of time, as with company profits, blocks are drawn instead of points, as in Figure 2. If the number of measurements is large (say, more than twelve), it is possible to use dots joined by straight lines instead but, in general, blocks are preferable. The time scale is always the horizontal one.

Sometimes the variation in the quantity is small compared with its general size, as in Figure 1 where the variation is less than 10% of the totals. When this happens, a scale starting at zero would make it difficult to follow the changes accurately, and it is quite legitimate to use a scale starting at any convenient point, *provided* that attention is drawn to the fact. This is done by showing a dotted or zigzag section at the foot of each vertical line.

If, on the other hand, the intention was to show how small in proportion the variations were, the complete scale would be used—an example of how the choice of scales can affect,

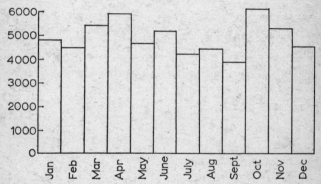

Figure 2   Time series of the total quantities in each successive time interval

quite fairly, the impression given by a graph. Careful choice of the starting-point on the time scale can also, perhaps less legitimately, alter the appearance quite a lot.

Two or more quantities are often plotted on the same graph, either for purposes of comparison or to save space. An example is to show average incomes, both before and after tax, over a period of time. The vertical separation between the two lines shows the amount of tax, though variations of this amount are not easy to follow accurately. The opportunities for distortion, deliberate or otherwise, are much greater than in graphs of a single quantity.

It hardly needs pointing out that a graph without units on its scales is valueless. Examples are few, as indeed they should be, but a choice example (reproduced below) appeared in an advertisement issued by the Milk Marketing Board. It was designed to show that the price of milk had risen less than the cost of living during the same eighteen-

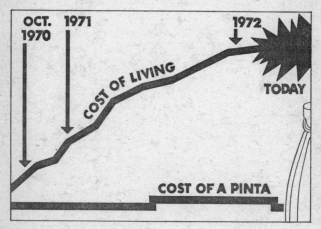

Figure 3 A misleading graph

month period—as in fact it had, the change being 10% compared with a 13% rise in the Retail Price Index over the same period. But the price levels were unscaled, and the change in the Index was exaggerated by a factor of about 15 compared with the change in the price of milk.

Various methods are now used for graph-drawing by computer. There is nothing new in principle about the actual graphs, except that sometimes the vertical measure has to be rounded off in fairly large units because the computer output has to be printed line by line. This gives the graph a stepped appearance which can be slightly misleading.

Logarithmic scales

The use of logarithmic scales is becoming increasingly popular, even in graphs designed for the general reader. It is a trend much to be encouraged, as the advantages in interpretation are considerable. It is not necessary to understand anything about logarithms in order to see how the scale works.

Consider a quantity which starts at 100 and increases at a constant *percentage* rate of, say, 20% per year. After one year the value is 120, two years 144, three years 172·8, four years 207·36 and so on. The numerical increases each year are 20, 24, 28·8 and 34·56 respectively, which themselves go up each time. Thus the graph against time of a quantity which is increasing at a fixed percentage rate is not a straight line but a curve. It is known as the exponential curve, and is observed in a great variety of practical, physical and financial measurements.

But a curve is less easily judged by the eye than a straight line. A series of numbers such as 100, 120, 142, 166, 192 would show an upwards curve similar to that of the exponential, but the percentage rate of increase is in fact going down all the time. The logarithmic scale is a way of distorting the graduations on the vertical axis so that the graph of a quantity which is changing at a constant percentage rate appears as a straight line, as in Figure 4. Variations from a constant percentage correspondingly appear as variations from the straight line, and are clear at a glance.

A quantity which increases at a fixed rate will double itself over and over again in equal periods of time—in the case of a 20% increase, the period is about 3·8 years. The method of constructing the scale may be seen most easily by noting the points at which this happens: 100 to 200 after 3·8 years, to 400 after 7·6 years, to 800 after 11·4 years. Since these are equal periods of time and since the graph is to be straight,

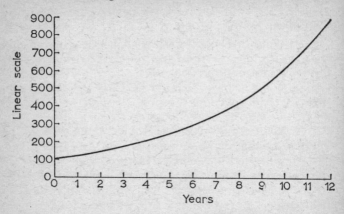

Figure 4 Time graph of a quantity increasing at a constant rate of 20% per year

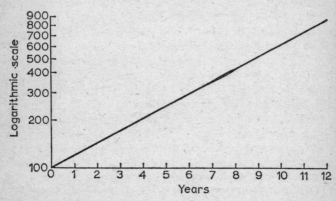

Figure 5 The same information as in Figure 4, shown with a logarithmic vertical scale

the vertical scale will have equal lengths to cover 100 to 200, 200 to 400, and so on. This may be checked with a ruler. Insertion of the other points requires some calculation, but these few figures should be sufficient to see roughly how the system operates. The scale lengths from 300 to 600 and from 500 to 1000 are equal to that from 100 to 200; other equal percentage increases, such as those from 400 to 600 and 600 to 900, are also represented by equal lengths.

The reason for the name is that, if the numbers on the scale are replaced by their logarithms, the scale required for the logarithms is a normal evenly divided one. Readers familiar with the slide rule have met a logarithmic scale already.

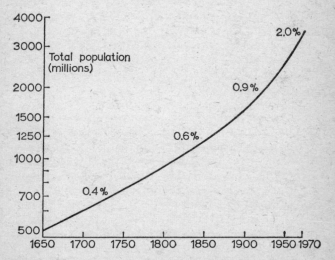

Figure 6 Total world population, 1650–1970, shown on a logarithmic scale. The increasing slope shows that the annual percentage increase (shown in figures also) has itself increased steadily

If a quantity decreases at a fixed percentage rate, the logarithmic scale is again appropriate, and the graph will of course have a downwards slope. But under these conditions a measurement can never get down to zero—consider a quantity which is halved every year, for example. So a logarithmic scale can never include zero.

The commonest example of exponential growth is a human or animal population. Recently the percentage growth rate of the world population has itself been increasing so that, even with a logarithmic scale, the graph shows an upwards curve, as in Figure 6. (This particular growth pattern is discussed in more detail in Chapter 16.) Energy requirements, industrial output, use of raw materials, company profits, inflation—all these are followed better by a logarithmic than by a linear (evenly divided) scale. Similarly, the decreasing true value of a unit of currency during a period of inflation is shown better on such a scale: a good example of a measurement which can never, we hope, reach zero.

The histogram

The histogram is similar in appearance to the time graph, but quite different in principle. It shows in graphical form the effect of the grouping process described in the last chapter. An example is shown in Figure 7. The horizontal axis is marked in the units of the data, showing the class divisions. In the simplest form, the number of observations in each group is shown by a block whose height is proportional to that number. Sometimes it is helpful to use more than one size of class interval on a single graph, so that the height is no longer a true measure of the number in the group; in this case the area of the block is used instead. In principle, it is better to use the area consistently, but it is an advantage to be able to use a vertical scale. If the number of groups is large, the heights may be shown by dots which are then joined by straight lines, making a form of graph called a frequency polygon. But the block form is generally preferable. A further variation is to join the dots by a smooth curve, which is a suitable form for quick approximations as it is very easily drawn.

A useful technical term may conveniently be introduced at this point. The **distribution** of a statistical quantity is a term which implies that the relative numbers of each possible measurement are being considered. A histogram is therefore a graph of the distribution of a quantity of which a number of different measurements have been made. It shows at once whether the distribution is symmetrical or not, regular or irregular, humped or fairly uniform—in other words, it is the most convenient way of giving a simple description of the

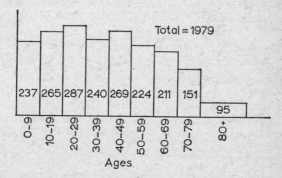

Figure 7 Histogram showing the age distribution of the population of Middletown on April 25th, 1971

distribution. From it, an estimate may be made of the mean, and if necessary the standard deviation, of the distribution, thus giving a rough check on the accuracy of any calculations which may later be made.

The histogram may also be used for showing in graphical

form the results of statistical experiments like dice-throwing or coin-spinning. If five coins were thrown together 1000 times, the frequencies of the six possible results (0 to 5 heads inclusive) could be plotted against the number of heads. A further example of a histogram, the age pyramid, is mentioned in Chapter 16.

A variation known as the cumulative frequency curve is sometimes used. Here the vertical scale is the total, or cumulative, number of observations up to that point, and the horizontal scale is as before. The points on the graph may be joined by straight lines or, preferably, by a smooth curve. An example is shown in Figure 8. The median salary (that of the man half-way up the list in order of size) is about £1800, and about 170 of the 200 men have salaries below £3000.

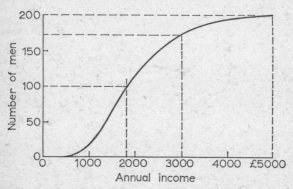

Figure 8   Cumulative frequency curve for the incomes of a sample of 200 men

The general form of a cumulative frequency curve is shown in Figure 9. It will be seen that the quartiles may be estimated in a manner similar to that for the median. The range and the quartile deviation (mentioned in Chapter 4) may also be found as indicated.

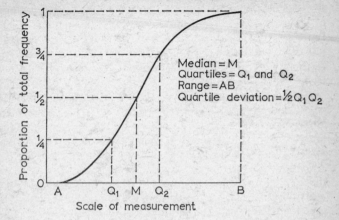

Figure 9 Cumulative frequency curve showing median and quartiles

Displaying data by categories

Many types of statistical data are best classified in categories, rather than by the values of one particular measurement as in the histogram. Such information is often shown pictorially, and three methods of doing this are given here: the bar diagram, the pie chart and the pictogram. Figures 10 and 11 represent the same data.

The bar diagram (Figure 10) is similar in appearance to the histogram, although the bars are often shown with gaps between them. Its main use is to show the relationships of the various parts to each other rather than to the whole. The bars may be shown either vertically, as here, or horizontally. Examples in which this type of diagram can be used are numbers of migrants classified by countries of origin or destination, or membership of various trade unions within an industry. Sometimes a single bar, divided into sections, is

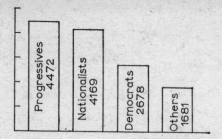

Figure 10   Bar diagram showing election result in Barsetshire North

used, for example to show how each pound raised by taxation is shared between health, education and so on. But comparisons between categories are then not so easy.

The pie chart (Figure 11) is so called because it looks like a pie divided into sections. Here the immediate visual effect is

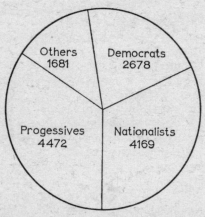

Figure 11   Pie chart showing election result in Barsetshire North

to show the relationship of each part to the whole: if one category accounts for about a quarter of the total, this is clear at a glance. Comparing the relationship of one category to another is more difficult, particularly if they are on opposite sides of the pie. The number of categories must be fairly small to avoid congestion, and categories accounting for less than 2% of the total are difficult to show clearly.

Twenty viewers tuned in

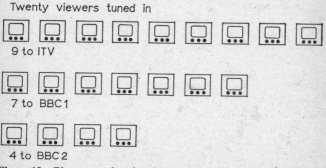

9 to ITV

7 to BBC1

4 to BBC2

Figure 12 Pictogram showing the proportion of viewers tuned to three television channels at a given time

The pictogram (Figure 12) is a variation of the bar chart, used for added visual attraction. It is most used in printed work where the symbols can be reproduced with ease; in manuscript the drawing process can be tedious. As a further variation, pictures of different sizes can be used, but care is necessary as the human eye naturally compares areas rather than lengths of lines. For example, a square of side 2 cm *looks* about four times the size of a square of side 1 cm— certainly more than twice as big.

The scatter diagram

When two separate measurements are thought to have some degree of association, a graphical method of presenting the information can serve to indicate in very general terms how close the association is. To take a simple example to start with, suppose that oranges cost 4p each, so that there is an exact relationship between the number bought and their total cost. We can show the pairs of observations as in Figure 13, and it will be seen that all the points lie on a straight line which passes through the zero point on the axes, called the origin.

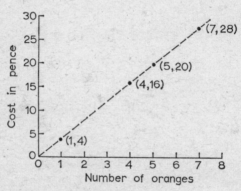

Figure 13   Linear relationship

Now suppose that various selections of oranges are bought at assorted prices, some less than 4p and some 4p or more, and that the results are shown in the same form as before. The pattern will be of the form shown in Figure 14.

The relationship is no longer exact, but there remains some degree of association, which is expressed in general terms by saying that a large number of oranges cost more than a small number.

Such a graph is called a scatter diagram, and the closeness of the points to a straight line drawn through the middle of them is a measure of the degree of association or correlation. As a further example, suppose that fifty men were asked to

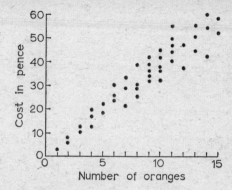

Figure 14 Scatter diagram

state their heights and weights. Then the scales for the two axes would be chosen to cover the ranges of height and weight respectively, and some correlation would be observed. If the men were then asked to state also their salaries, a scatter diagram showing heights and salaries would probably show no particular pattern, since there is no obvious reason for supposing that salaries of tall men are in general any different from those of short men.

Sometimes there is an inverse association between two measurements—when one goes up the other goes down. For instance, the countries of the world which have higher than average *per capita* incomes tend also to have lower than average birth rates. Here a scatter diagram has the points grouped around a line which slopes downwards rather than upwards, as shown in Figure 15. (The number of sizeable countries with high birth rates and with both quantities known with reasonable accuracy is unfortunately rather small.)

The scales on a scatter diagram are not important, in the sense that the degree of correlation indicated by the graph is

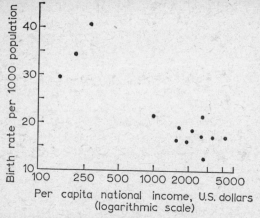

Figure 15   Scatter diagram, incomes compared with birth rates (countries of over 10 million population; data of 1970)

independent of the scales chosen. The range on each scale need therefore be no longer than the range of the actual measurements.

To judge the actual degree of association, and to estimate whether it is genuine or might be just due to coincidence, requires some skill. In general, the more observations there are, as well as the clearer the pattern, the more certain one can be that an association really exists.

But proof of association is not the same thing as proof of cause and effect, which is usually a matter for other than purely statistical techniques. The high correlation between smoking and lung cancer might as easily be used to prove that incipient lung cancer makes people start smoking as that smoking causes lung cancer. When this objection was first raised, further investigations were made in order to test it, as described in Chapter 12; the possibility has now been discounted.

6 Probability

Most people acquire a working knowledge of probability through contact with games of various kinds—spinning a coin to give the choice of start in a team game, dice games from snakes and ladders to poker dice, card games, roulette and gambling games in general. We also use the concept of probability in the ordinary events of life, notably in assessing weather prospects. In simple cases, the scale of measurement can be worked out intuitively, for example in saying that the probability of throwing a double-six with two dice is less than that of throwing a six with a single die. But more often some mathematics is required, for instance in poker dice or card games. In many situations only a subjective estimate of the probability level can be given, as in predicting weather or the result of a race.

The use of probability in statistical work is so frequent that it is essential for a full understanding of the subject. The same distinction arises as described above: in some situations it is possible to use mathematical analysis, and in others only subjective estimates can be made. The analysis can of course become very complex, and further problems appear when analytical and subjective methods are used together. But some elementary ideas should prove useful, starting with a look at the fundamentals of the subject.

The universe in which we live is governed by various physical laws, which can be used to work out the answers to all sorts of problems: what happens to an electrical circuit when a voltage is applied, how a violin string is tuned to give a certain note, when and where an eclipse will take place, and so on. So it should be possible to work out the answer to this simple little problem: if a coin is flicked upwards so that it falls to rest on a hard surface, which side of the coin will remain showing?

Of course, no answer based only on physical laws can be given. This is not because we do not understand the theory, but because there are too many factors affecting the answer, most of them too complicated to measure except in laboratory conditions. Also, a tiny alteration in any one of them can change the result completely.

Probability deals with this sort of situation. It is an attempt to set down in a quantitative manner what is observed to happen in practice, in situations like coin-spinning where the physical and mathematical laws of nature cannot give the answer.

Another idea, that of symmetry, may be used alongside probability in order to make interpretation of the process easier. (However, the concept of probability is not in any way dependent on symmetry.) A coin is for practical purposes symmetrical as between head and tail—that is, there is no feature relevant to the coin-spinning experiment which distinguishes the head from the tail. Hence we can say that the probability that the face falling uppermost is a head is 'an even chance', or any other expression meaning the same thing, such as '50-50', or '1 in 2'. This intuitive approach suggests the numerical value for the probability in this instance, 1 in 2 or simply $\frac{1}{2}$.

By similar reasoning, the probability of each of the six possible results when a die is thrown is $\frac{1}{6}$. When two dice are thrown together, there are thirty-six different ways in which they can fall, assuming that, say, 4–3 is counted as different from 3–4, because each of the six possible scores on the first can be combined in turn with each of the six for the second. Since conditions of symmetry still obtain, the probability of each different pairing is $\frac{1}{36}$.

But what do all these actually mean in practice? For the coin, simply this: the probability value of $\frac{1}{2}$ implies that in a very long run of spins of a fair coin, a proportion of about a half would show heads and the remainder tails. Similarly, for a die, about one-sixth of the total number in a long series of throws would be ones, one-sixth twos and so on. In, say, 36 000 throws of two dice, about 1000 double-sixes would be expected, since the probability is $\frac{1}{36}$.

These results are essentially statements of experience, not of a theory which is open to mathematical or experimental proof. Again as a matter of experience, we can say that the longer the run, the closer (generally speaking) the proportion will get to the exact value predicted from symmetry. The numerical value of a probability, therefore, may be defined as the value this proportion would reach if the operation were to be repeated an unimaginably large number of times. Such a value is called by mathematicians the *limit* of the proportion.

We ought not really to call this a definition, because it begs the question by supposing that such a limit exists. It is perhaps better described as an axiom: it is in accordance with common sense, it works in practice and there appears to be no way of giving a direct proof of it.

Although the examples used so far have involved symmetry, the definition applies equally in situations where there is no symmetry. Suppose we have a coin which is slightly bent. Common sense suggests that one face will fall uppermost with a higher probability than before: that is, the proportion of throws showing that face will increase to a value measurably greater than $\frac{1}{2}$. But this time there is no way of predicting the actual value of the probability, and an estimate has to be made by counting the numbers of heads and tails in a long run of throws: the longer the run, the better the estimate.

Similarly, if birth records show that about 515 out of every 1000 live births are male, then the probability that the next child which will be conceived by a particular woman is a boy is 0·515. (The figure given here is approximately correct and, although the proportion varies a little from place to place, it is consistently over $\frac{1}{2}$. Notice the rather careful phrasing of the statement, by the way: if the baby is already in the womb, one cannot state a value for the probability as the sex has been determined in fact, even though so far unknown to anyone.)

The measure for probability therefore goes on a scale from 0 to 1. Zero represents impossibility, the probability that the sum of the scores on two dice is 13; 1 represents certainty, the probability that a double-headed coin will show a head after spinning.

It follows from the proportion definition of probability that it must be possible to repeat an experiment any number of times, if a numerical value for the probability is to be found. In the repeats, there must be no changes which might affect the probability level. We cannot therefore use the concept at all in non-repeatable situations, which means that we cannot make such a statement as, 'The probability that Gambler's Folly will win the Derby is $\frac{1}{3}$'. If we want to do so, we must use a different definition altogether—one which considers probability in the sense of being a degree of rational belief. Such a definition is used in some kinds of statistical work, but the general reader is most unlikely to come across any mention of it. The methods based on it are in the nature of being rather second best, to be used only when others do not work, because the logical basis is a good deal weaker than that of the proportion definition.

Every time a numerical value of a probability is used, it is necessary that the requirement mentioned earlier should be rigorously observed: in the sequence of repeated operations there must be no change which might affect the probability level. In principle, this is obvious enough. In a race between ten horses, for example, it is not true that each has a probability $\frac{1}{10}$ of winning, because it is known from previous experience that some are faster than others. But often the fallacy will not be so evident.

Suppose it is known that 95% of first-class letters are delivered on the day after posting, and that I wish to know the probability that a letter which I am about to post will be delivered tomorrow. In this case, once again, no value for the probability can be given: it may be anything from zero (if the address is far distant) to almost 1 in the case of a local delivery. The millions of letters sent each day are not repetitions of each other, in the sense that large numbers of throws of a coin or a die are repetitions.

It may be thought that this requirement appears to restrict the use of probability to artificial experiments which use coins and dice, rather than to situations involving actual people who are by no means carbon copies of each other, but this is not so in fact. The reason cannot be put into a few words, but it should become clearer after reading the next chapter. The example below may also help.

In the 1961 Census of the United Kingdom, one household in ten was given a longer list of questions to answer than the rest, and the results were used to work out (among other things) the number of people with certain scientific qualifications. It was assumed that the number found in the fraction of the population actually asked the question could be scaled up, without serious error, to give an estimate of the total number. The line of reasoning is just the same as if coins or dice were involved—that the probability that one individual, chosen by lot from the whole group, will turn out to be a scientist is the same whether the group consists of the country's whole population or only a tenth of it, and so the number actually found within that tenth can be scaled up to give the total. But the choosing of the one household in ten has to be done with some care, as will be seen in the next chapter.

A simple experiment which is similar in principle can be used as a test of the extent of the error involved in the scaling-up process. The mathematical aspect of the problem is not discussed, however, so that no justification can be given for the figures which will be quoted. Readers who wish to make good the omission will have to consult a more advanced book. The experiment is rather like coin-spinning, but instead of a coin an ordinary drawing pin is thrown onto a hard surface. The result is recorded as a 'point' if it falls point-upwards, and a 'head' if it comes to rest with the point and part of the head touching the surface. The probability of a head may be estimated by throwing the pin a large number of times, or by throwing several at once if they are identical. When this was done by the author, 200 throws produced 85 heads, a proportion of 0·425.

Such a figure is very little use in trying to predict the result of the next single throw. But for large numbers of throws quite high accuracy is possible, and the degree of accuracy itself can be estimated on a probability basis. One could not *guarantee* that the number of heads in the first 2000 throws would be in the range, say, 700 to 1000, that is within 150 each side of the scaled-up value of 850. But the probability level can be stated and is, in fact, about 0·98. The 2000 throws actually produced 794 heads, as it turned out.

The larger the numbers, the more accurate the predictions become. Given 794 heads in 2000 throws, the proportion in 20 000 throws could be predicted to within 3% with a high degree of confidence. For the number of scientists in the example quoted above, even higher accuracy was to be expected, as the numbers were larger still.

This, essentially, is what is known as statistical inference: the art of drawing conclusions about a large number of people or things by examining or testing a section of them only. Public opinion, to take one of the most obvious examples, is estimated by researchers who ask, not 30 million people, but about 3000 at the most. As with the drawing pins, the conclusions reached by such a method will not be totally accurate but, if certain conditions are observed in the selection of the sample, the degree of accuracy can be estimated. The conditions themselves are the subject of the next chapter, but it is appropriate to mention here the way in which the conclusions will be expressed. It is not a particularly easy subject to explain without advanced mathematics, and may be omitted if desired.

Suppose we have an experiment in which the probability of event is shown by lengthy tests to be $\frac{1}{2}$. (This value is chosen in order to keep the arithmetic simple, not because the results depend upon it.) A tossed coin is the obvious example, but the symmetry of the coin once again has no bearing on the situation. We throw the coin twenty times. About ten heads may be expected, but what we really want to know is this: how are we to estimate the likelihood of a result in the range, say, 9 to 11, or 7 to 13? The latter is clearly more likely, but how much more?

If we have one coin, the probabilities are $\frac{1}{2}$ for a head (H) and $\frac{1}{2}$ for a tail (T). Now suppose that the coin is thrown twice. Four results are possible: TT, TH, HT and HH, with probabilities $\frac{1}{4}$ for each. These can be summarised as $\frac{1}{4}$ for no heads, $\frac{2}{4}$ for one head and $\frac{1}{4}$ for two heads. For three throws, there are eight possible results: TTT, TTH, THT, THH, HTT, HTH, HHT and HHH, giving probabilities $\frac{1}{8}$ for no heads, $\frac{3}{8}$ for one head, $\frac{3}{8}$ for two heads and $\frac{1}{8}$ for three heads. If you find this difficult to follow, notice that the first four results consist of a T followed by each of the four two-coin combinations in turn, and the remaining four consist of an H followed by the same group.

We can now start to build up a table of probabilities, as follows.

Number of throws	Number of heads					
	0	1	2	3	4	5
1	$\frac{1}{2}$	$\frac{1}{2}$				
2	$\frac{1}{4}$	$\frac{2}{4}$	$\frac{1}{4}$			
3	$\frac{1}{8}$	$\frac{3}{8}$	$\frac{3}{8}$	$\frac{1}{8}$		
4	$\frac{1}{16}$	$\frac{4}{16}$	$\frac{6}{16}$	$\frac{4}{16}$	$\frac{1}{16}$	
5	$\frac{1}{32}$	$\frac{5}{32}$	$\frac{10}{32}$	$\frac{10}{23}$	$\frac{5}{32}$	$\frac{1}{32}$

The fourth and fifth lines can be checked using the same method as before. Thus the number of combinations of head and tail in four throws is 16, written down by combining first a T and then an H with each of the three-coin possibilities in turn. But a simpler method of building up the table can also be seen, provided that no cancelling of fractions is done.

There is a pattern in the denominators (lower parts) of the fractions which is clear at once: they double each time, working downwards. There is also a pattern in the numerators (upper parts), which is that each is the sum of two numerators in the line above, the one immediately above and the next one to it on the left. Thus the line for four throws is found in two stages. First the denominator is found by doubling the 8 from the line above. Then the first numerator is 1, the second 1 plus 3, then 3 plus 3, 3 plus 1 and finally 1 again. (The 1's fit into the system if a blank space is counted as zero.) The line for 6, similarly, has denominator 64 all through, and numerators respectively 1, 6, 15, 20, 15, 6 and 1. As a check in each line, the sum of all the numerators should equal the denominator, since the sum of all the fractions must be 1.

The process may now be continued, the only limitation being that the numbers soon become unmanageable. The probabilities of 9 and 10 heads out of 20 are respectively $\frac{167\,960}{1\,048\,576}$ and $\frac{184\,756}{1\,048\,576}$; that for 11 is the same as that for 9, by symmetry. So the probability of getting either 9, 10 or 11 heads out of 20 is the sum of these three, $\frac{520\,676}{1\,048\,576}$, or just under 0·5.

So, if we wish to predict the result with a strong probability of being right, wider limits than 9 to 11 will have to be chosen. (A weather forecaster who got his predictions right less than 50% of the time would soon find himself out of a job.) By altering the range to 7–13, the proportion goes up to 0·885, so that a prediction that the number of heads would be within these limits stands a better chance of being right—8 times out of 9 on average.

We can thus give a range of values, rather as we give a tolerance. But it is not certain that an actual value will lie inside the range quoted, and in fact the probability that it will not do so may be stated. We can choose the range to suit any given probability level, such as 0·95 or 0·99; the higher the probability, the wider the range which must be allowed.

The form of conclusion which may be drawn from sample measurements follows approximately along these lines. Suppose, for example, that out of a properly selected sample of 400 voters 208 declared that they would vote Labour in an election which was to be held. This gives a proportion of 0·52 but, because the sample is rather small, the error may be greater than the 0·02 required to alter the predicted result. In fact, the necessary calculations (not explained here) show that the possible range of values is 0·52 ± 0·02, at a confidence level of 95%. Such a 'confidence level' is rather like the probability measure described above, though there is a technical distinction which need not concern us here. The implication of the conclusion is therefore that we can be reasonably certain that the proportion of Labour voters in the whole electorate is in the range 0·50 to 0·54, but that there is just a small chance (5%) that the true proportion might lie outside those limits.

One further point requires explanation. With twenty coins,

the probabilities of the various possible results were already becoming awkward to calculate because of the large numbers involved; with 400 voters, a new method is clearly required. The answer lies in a piece of mathematical analysis

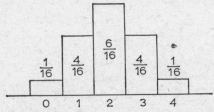

Figure 16　Coin-spinning: probabilities for each possible number of heads in 4 throws

which is of far-reaching importance in statistics. The underlying process can be seen very clearly by drawing a graph of the probabilities of each possible result in the coin-spinning experiment, for several different numbers of throws.

The set of probabilities in each case is called a probability distribution, by analogy with a distribution of observed results as described earlier. Each distribution has a mean, which by symmetry is half the number of throws, and also a standard deviation. The interesting point about the distributions

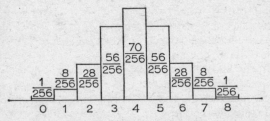

Figure 17　Probabilities for the possible results of 8 throws

is that they all have roughly the same shape, looking at the general outline rather than at the individual blocks: a central hump with a tail on each side. (If a probability level other than $\frac{1}{2}$ is chosen, the general shape is the same, but

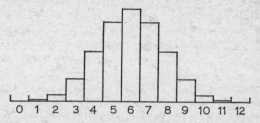

Figure 18 Probabilities for the possible results of 12 throws

there is some degree of asymmetry which gradually becomes less as the number of repetitions increases.) With increased numbers the small errors introduced by drawing a curve instead of the series of blocks become even smaller; a further

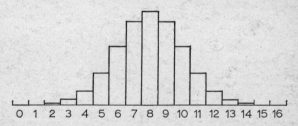

Figure 19 Probabilities for the possible results of 16 throws

adjustment known as a continuity correction can also be used to allow for these differences.

So the final result is a single curve which, after adjustment of the scale, can be used to calculate probabilities for any number of throws. The probabilities can be tabulated, like tables of logarithms, so that the solution of a particular problem can be found in a very short time.

As briefly indicated above, the probabilities need not be $\frac{1}{2}$ for this same curve to be used. (The mathematical proof of this is rather difficult, and there are certain limitations on the use of probabilities close to 0 or 1.) The curve turns up as well in all sorts of other situations, sometimes simply as a matter of observation and sometimes when theoretical justification is possible. Perhaps the most useful of all these is the one mentioned earlier—the estimation of the degree of confidence which can be attached to certain results obtained from samples. The curve is called the **normal probability distribution,** and it has the great merit that, in the situations in which it can be used, it enables precise statements about probability levels to be made without lengthy computation.

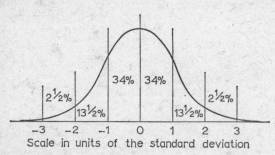

Figure 20 The normal distribution

As mentioned in Chapter 4, the fixed shape of the curve means that the percentiles which correspond with each point of the distribution can be worked out in terms of the standard deviation. Some of the results are shown in Figure 20.

This, once again, is about as far as one can go without mathematical analysis and the detailed use of the standard deviation.

7 Taking samples

The process of reaching conclusions by taking samples is basic to each one of us—we started it in the first year of our lives when a sample of one mouthful of a particular type of baby food made us decide that a second mouthful was either to be taken eagerly or to be rejected. A simple justification of the method, based on the laws of probability, was given in the last chapter. But this justification is dependent on the sample being truly representative of the group from which it is drawn. In principle, this is only common sense: no market researcher or opinion tester would carry out his investigations in an area known to have an unusually high proportion of elderly people or business executives, and then apply his figures to the country as a whole. But many of the pitfalls are better disguised than these.

Sampling may be made necessary for one of three reasons:

(i) The testing process may be destructive so that, if all the population were tested, nothing would be left. (The word 'population' is used here in a technical sense to mean the whole group under consideration, not necessarily people.) Wine-tasting is one example, the wine being the population.

(ii) The population may be too large to test as a whole, for reasons of either cost or time. Examples are many types of manufacturing process, and all opinion testing. There may be advantages in sampling, even if the whole population could be tested, because smaller numbers means better control over the testing procedure: one example, the General Household Survey, was discussed in Chapter 2.

(iii) The population may be indefinable or infinite, as in the drawing-pin experiment where there was no limit to the possible number of throws.

The errors involved in drawing conclusions from samples are of two kinds. Some are unavoidable, created whenever sampling is used, and were discussed in the last chapter. The rest are avoidable, in principle at least, and are caused by using samples which are in some way unrepresentative of the population as a whole. The extent of the unavoidable error is, roughly speaking, determined by the size of the sample, and this in turn is determined by the amount of money available. The true art of sampling is to keep the avoidable errors to the very minimum in giving the answer to a particular problem, within the limits of the available time and, once again, money. Perfection in sampling is almost never to be attained; the work of a sample designer is the art of the possible. The methods which he can use are the subject of this chapter.

The sampling frame

The initial stage in planning a sampling procedure is to choose (or possibly construct) what is called a sampling frame—that is, a list or a description of the complete group from which the sample is to be drawn. For instance, it is no use starting to make an estimate of public opinion on some topic of the day without having a clear idea of the limits, both geographical and by age, of the 'public' one wants to test.

Suppose we want to take a 10% sample of all the primary school teachers in Birmingham. First, there are matters of definition to sort out: part-time teachers, private and special schools and so on, together with boundaries. Then we need some method of tracing the entire group. If some are left out—say, because they have recently moved to their present jobs—any sample drawn will be unrepresentative; in this case, teachers just out of college will be under-represented. The ideal is to have a numbered list, accurate when first written down and since that time kept up to date. It is not difficult in principle, but in practice is limited by time, cost and human fallibility.

If the whole population lies within known physical limits, and if the items required are readily distinguished from extraneous ones, there is no need to have a list of the items, although a different sampling method will have to be used. Articles coming along a production line form the commonest example of this type of frame.

A sampling frame presents most difficulties when it refers to people. The geographical boundaries have to be defined first, and for reasons which will appear later it is usually necessary to choose people on a basis of where they live, not on where they happen to be at a particular moment in time. There is no publicly available sampling frame for those under 18, and this is therefore the lowest convenient age limit for a sample survey. It may not be the limit best suited to the investigation being made, of course, but the researcher has to learn to live with restrictions of this kind.

The most convenient sampling frame for the whole adult population is the Electoral Register, which is revised every twelve months. Because of the time needed for preparation and publication, however, it may be up to sixteen months out of date at the time it is used, so that changes due to deaths and removals will have taken place. Some categories of person not qualified to vote are not included, and other names are left out simply because people fail to supply the information needed. (There is a legal requirement to register, but in practice no complete check is made.)

The same imperfections, unfortunately, exist in other frames as well. Street directories and rating lists soon go out of date, and also contain inaccuracies created by the manner of their construction. The Inland Revenue and National Insurance departments have their own lists, but the information is confidential and for various reasons is again not complete. The same is true of the registers of people who receive retirement pensions and families entitled to child allowances.

None of these sampling frames is made up for the purpose of statistical work, and therefore they do not necessarily have the information which would be most useful. The Electoral Register does not give age or marital status, and even a person's sex has to be inferred from the name itself. A wider range of information can be obtained from census returns, either from the published results or by special request, but only in forms which preserve the confidential nature of the material.

For more restricted groups, all sorts of different frames exist. Those who follow certain occupations are listed with fair accuracy in publications such as *Crockford's Clerical Directory*. But in other apparently similar cases the list may only imply membership of the corresponding professional association, which is not always a prerequisite for doing the job. Also, such lists do not always distinguish the practising from the retired, and deaths may pass unnoticed when life members have no subscription to renew.

Organisations such as registered companies and charitable groups are also listed officially, though a legal requirement to register is no guarantee of accuracy. Once again, information which would enable such lists to be used as sampling frames, particularly an indicator of size, is not always there. As well as the obvious difficulties this creates, it can make the construction of a stratified sample (explained later) very much more complex. There is no list of private firms and businesses which is anywhere near complete, though work on one is in progress. Trade directories vary in quality from the nearly complete to those made up of firms which choose to pay for their names to be included, which are useless as sampling frames.

An ideal sampling frame should be complete and up to date, with no duplicates and no items which do not exist; the choice of whether each item is to be included should be objective; each item should be described adequately; and the information should be readily available, preferably in a form suitable for computer input. But such an ideal is far from being realised in any available frame, and indeed could only be realised in a society in which individual freedom was severely restricted. Total information means something like total power to those who hold it, and a form of society designed to provide good sampling frames for its statisticians would not be pleasant to live in.

Selecting the sample

In choosing from the frame, one by one, the items which are to form the sample, the basic principle is that each member of the frame should have an equal chance of selection. This, in turn, implies that no human whim or preference may be allowed any part whatever.

The simplest kind of sampling procedure uses a probability experiment such as coin-spinning. Suppose we require a sample about $\frac{1}{16}$ the size of the frame. Four coins are thrown, giving a probability of $\frac{1}{16}$ that all are heads. If this does happen, item 1 of the frame goes into the sample. Repeat for item 2, and so on. The sample will not contain exactly $\frac{1}{16}$ of the number in the frame, but this will not matter for most purposes.

A different approach is to use the lottery method. If there are 50 000 items in the population, a corresponding set of 50 000 numbered but otherwise identical cards is placed in a container and shaken vigorously. Cards are then drawn unseen, one by one, until the sample is complete. There is a technical distinction between the method described here and the variation of it in which each card drawn is replaced after the number has been noted, but for most purposes the difference is small.

Both these methods are, as the reader has no doubt been thinking, highly impractical for a large population. In practice there is a short cut, a method which may be adapted to suit either a one-by-one or a lottery approach. What is required is a way of choosing a number from any given range, such as 1 to 16 or 1 to 50 000, so that every member of the range has an equal chance of selection. To do this with only pencil and paper, lists of random digits are used. These are large numbers of digits (one book is published which contains a million of them), which are generated by a probability method so that every digit in turn has an equal chance of taking each of the ten possible values. A selection from one list is:

> 5150 2240 0737 9645 0727
> 1151 7570 0358 5601 4496

Suppose we require a series of numbers in the range 1 to 50 000, as a simulation of the lottery method. Since the top limit has five digits, the first five are taken from the list, which gives 51 502. This is ignored because it is over the limit. The next five digits give 24 007, so the item with this number forms the first member of the sample. The next is 37 964, then (after a gap) 11 517, and so on until the sample is complete. A duplicate would in practice be passed over.

All sorts of variations are possible. If a series of random numbers in the range 1 to 16 is required (to save throwing four coins), it would be wasteful to take digits two at a time and ignore combinations over 16. Instead, the first number of each pair could be counted as 1 if odd and 0 if even, and the pairs can then be read as 11, 10, 02, 00, 07, 17, 16 and so on—only a few of which have to be ignored. Alternatively, an even digit may be read as a head and an odd as a tail, so that 2240 counts as four heads.

The best-known use of random digits is in the selection of Premium Bonds, a form of investment in which the interest payments each month are distributed on a basis similar to that of a lottery. The ten-digit random numbers from which the prize lists are made up are generated by ten pairs of 'noise-producing diodes'. When a current flows through each diode, a random movement of electrons takes place, which produces minute variations in the output. These variations, in the form of minute pulses of energy, are counted by an electronic method. One thousand times in each minute, the numbers on each pair of diodes are read automatically and added together, and the last digit of the result forms one digit of the ten-digit serial number. (Two of the ten are in fact letters, and therefore have more than ten possible values.) These serial numbers are recorded on magnetic tape for subsequent processing by computer.

Some of the numbers generated have to be discarded, either because the bonds have been redeemed or because they have not yet been sold; the filtering process is partly manual and partly automatic, with elaborate checks to ensure accuracy. An independent check of the randomness of the results is made each month, and has always confirmed the reliability of the method.

Readers who wish to try out the idea of using random digits but have no ready-made set available can manage quite well with a telephone directory. Not all the digits can be used, of course, as the area codes upset the balance, as well as other less obvious sources of bias. But the sequence of last-but-one digits of each number will be found quite satisfactory. A lengthier alternative is to throw a coin and a die, with a suitable arrangement for numbering ten of the possible results and ignoring the other two.

Those who are tempted to take a short cut by just writing down a list of numbers 'at random' are invited to try. A list of 500, written down in sequence, will be found sufficient to test the success of the attempt. All sorts of tests are possible, of which two are suggested here. First, count up the number of pairs such as 55, a triple to count as two pairs—there should be about fifty of them. Then count the number of groups of six consecutive even numbers—there should be about eight. If the list is used in the same order in which it was written down, a genuine probability model is, in the literal sense of the word, impossible to imitate.

The possibility of human control or interference nullifies the essential feature mentioned at the start—that a pure probability method is the only way of ensuring that each item has an equal chance of selection. The technical name for a sampling system in which this condition obtains is **random sampling**. In spite of the use of the word 'random' in everyday speech to mean something rather similar, it should not be used in statistics unless its use is justified in the sense specified, as in the phrase 'random digits'. In the sampling method described above, therefore, the phrase may be used but, in the variations of the basic method which are to be explained next, there is either some degree of non-randomness in the sampling or an assumption about the degree of randomness in the sampling frame, which amounts to the same thing. The word 'random' is often still used in such circumstances, but strictly it ought not to be.

The operation may be simplified in the situation in which the population is ordered but not listed, as with articles coming off a production line. Suppose that a 10% sample is required. A random digit is used to decide which of the first ten items should be in the sample, and thereafter every tenth item is chosen. The name given to this method is **systematic sampling**, which seems to belie the fact that selection is made solely on a probability basis: an alternative is to refer to it as 'fixed interval after random start', which describes the process exactly. Before the random digit is known, every item in the population has a probability of $\frac{1}{10}$ of being chosen, and the only difference of principle from the coin-spinning method is that the whole sample depends on a single random choice instead of many. The method is widely employed.

But it cannot guarantee a genuinely random sample, because there may be some non-randomness (that is, a pattern) in the ordering of the population. If every tenth item coming down the production line is faulty, the sampling system has a good chance of missing them all. The frequency in the pattern does not need to be exactly the same as the frequency of the sampling method in order to upset the system —it is only necessary that the two should have a factor such as 2, 3 or 5 in common. For example, a 1 in 10 sample from an electoral register would be no use in a street of young families: each house would contain two voters, and the man's name would be first, so that every tenth name would yield a single-sex sample. Or consider a bottling machine which fills eight bottles at a time, so that a fault in unit number 3 could escape notice if a 1 in 10 sample always made its choice from units 2, 4, 6 and 8. If there is any possibility of such a pattern, a little ingenuity will usually solve the problem: in this case, a 1 in 9 system would suffice. As an alternative, take a random digit to start with, and then the 9th, 19th and 29th of each succeeding 29. This will yield a sample of just over 10%, which will be random unless there is a pattern in the population which has a frequency of some multiple of 29.

The method is frequently used in house-to-house surveys. The fixed interval chosen in this case must be an odd number if the houses forming the sampling frame have all the even numbers on the same side of the street. The interviewers must also be given careful instructions about the choice of the first house, preferably by choosing the number randomly beforehand. To start always at number 1 would give a bias towards corner houses. Different methods are required for rural areas, and for larger units such as hotels.

It is essential that no discretion should be allowed to the interviewer. In the 1961 Census of the United Kingdom, a 10% sample survey was carried out at the same time which showed a bias away from large families. The bias was well outside the usual sampling error and could only have been caused by enumerators failing to follow their instructions to the letter. To a person not statistically trained, the occasional slight variation no doubt seems a trivial matter, but the errors are biased rather than compensating and in this case the total effect was quite large—over 20% in some categories.

As a result of the experience gained in 1961, future surveys of this type will probably be arranged in such a way that an enumerator who is willing to break the rules to save himself trouble cannot cause bias of this kind. Two methods are possible. One is to ensure that the actual 1 in 10 selection, made on the basis of a random last digit of the census form serial number, is not revealed to enumerators until after the forms have been distributed. The other method, which is more difficult administratively but has much to commend it, is to use ten different forms in sequence, so that the type of form is identified by the last digit of the serial number. Questions to be answered by everyone then appear on each form, while questions to be answered by a 10% sample are divided up among the ten in such a way that all the types of form are of about equal length. A further advantage of this procedure is that the sampling fraction can be altered to two-tenths, three-tenths and so on for particular questions, instead of being fixed at one-tenth.

The 1971 method—to give identical questions to everyone, but in some cases only to analyse 1 in 10 of the results—is, by contrast, clumsy in the extreme and is most unlikely to be repeated.

An alternative way of taking a systematic sample is to choose an arbitrary feature possessed by some members of the population which bears no relationship to the purpose of the survey, and to use this as a way of deciding whether an item should be included in the sample or not. For example, if a large firm wished to ask a sample of its employees for its opinion of the canteen meals, those whose surnames began with D could be used to form the sample. There is no reason to think that this group might show a bias towards management rather than shop floor, or any other related to the purpose of the test. But there are both regional and national differences in the pattern of surnames, and in some circumstances this might be important. The method is called **attribute sampling**. Anyone who has the task of dividing twenty-two enthusiastic youngsters into two teams for football is recommended to try the idea: take a 'sample' by dividing off those with birthdays in the first six months of the year, and then adjust numbers as necessary—it can be done so quickly that protests of unfairness have no time to materialise.

A more serious example is the National Child Development Study in which all the 17 000 babies born in England, Scotland and Wales during a particular week in March 1958 have been used to provide the subjects for a continuous study of how children actually develop—the most far-reaching project of its kind ever undertaken in the United Kingdom. The sampling method assumes only that the qualities being investigated are independent of the month of birth.

If the population is very large, then for a sample of a particular size the sampling fraction of course becomes smaller: for an opinion survey in the United Kingdom a sample of 1 in 10 000 would be sufficient. To select the names from the electoral registers would be easy enough, but to visit them all would be time-consuming and uneconomic because they would be scattered all over the country. So in practice certain compromises have to be made. They all introduce either an element of subjective choice or an assumption about the type of population in a particular area, so that the concept of a true random sample is lost. The best one can do is to try to keep the resulting bias down to a minimum; in a well-designed system the errors may be small, but an important point of principle has been lost and, where criticisms are voiced, the statistician has to rely partly on subjective argument in order to answer them. To defend the results of a truly random method requires no reliance on information about the population being sampled, because it relies solely on the mathematics of probability.

The first type of compromise is the **stratified sample**. The method is to divide the population into homogeneous groups, as in the strata or layers of various types of rock. The groups may be of any kind provided that they are reasonably well defined and that the proportions of each group in the whole population are known fairly accurately. The socio-economic groups devised by the Registrar General and by the Institute of Practitioners in Advertising (see Chapter 9) may be used, for example. The numbers from each group required to form the sample are then calculated from the known proportions in the population. Samples drawn from a restricted part of the whole population, probably by a systematic method, are then arranged so as to fit these requirements. Information about the income levels in particular areas is available from census data, and this may be used for stratification; a further method is to use information about the industrial and domestic rateable values in the same way. But, whatever the method used, the ever-present economic constraints may still have an effect, making it, for instance, difficult to ensure the right balance between the urban and rural sections of the population.

For populations other than human ones, stratification may also be used and is sometimes the only possible method. When the units in the population differ widely in size, for example, serious distortion could occur if the proportion of very large units happened to turn out too big or too small in a random selection from all the units. Strictly speaking, the best course of action in these circumstances is to make the random choice in such a way that the probability for each unit is in proportion to its size instead of being the same for all. But a system of stratification by size is much easier to arrange and is quite satisfactory for most purposes.

A more drastic modification along these lines is called **quota sampling**. Here the groups chosen must be distinguishable by sight—age/sex groups, typically. The arrangement of numbers required for each group is worked out on a similar basis to that of stratified sampling. The interviewing may then be house-to-house but is more often in a public place, and continues until each quota has been filled. This is open to all sorts of errors, the chief one being that the people present in a public place are not representative of the population as a whole: those in full-time employment, particularly, are under-represented. Also, if age groups of 20/40 and 40/60 are used, an interviewer may be reluctant to approach someone who looks about 40 because she is not sure which category to put him in. So the sample may be biased towards those with ages in the central part of each group. Quota sampling is cheap, but its errors may in the long run prove expensive. It falls far short on the prime requirement of a sampling method—that no human whim or preference must play a part in it.

Another way of cutting costs is to take a **clustered sample**. The principle of this is the same as that of the basic random digit method, but a group of individuals is selected together rather than one at a time. So, instead of choosing 100 names from an electoral register, 10 are chosen, and for each one 9 more are selected near it, probably on a fixed-interval system. The effect is similar to that of using a rather smaller sample, assuming that there are variations in the type of population from place to place. If, for instance, one section of the population, 5% of the total, was in the case quoted different from all the rest, it would either get no groups out of the 10, or one or more. So there would be either no individuals out of 100, or 10 or more, and the proportion of this 5% group who are represented in the sample is no more accurate than if a sample of 10 individuals had been drawn. A sample of 100 separate individuals would be likely to produce a proportion closer to the true one.

Next there is a process called **multi-stage sampling**. Suppose that a sample of 1000 is required from a population of 100 000. The population is divided into 100 groups of 1000, and 10 of these groups are selected using random digits. A 10% sample is then taken from each of the 10 groups. With larger numbers still, three stages may be used.

Obviously, endless variations are possible. In practice, the groups must be based on some pre-existent structure, such as the administrative areas of the country, and so the numbers will not be as tidy as those given here. It is therefore necessary to make a correction for the fact that the areas are of different sizes, preferably by using ranges of random digits in such a way that the probability of choosing an area is in proportion to its size. Alternatively, a stratified system may be used. But there are further errors inherent in the method, very similar to those in cluster sampling.

All these methods have been based on an assumption that the sampling is to be a one-off operation. For opinion sampling, it is often of interest to be able to assess the movements of opinion over a period of time, and to do this it can be an advantage to use the same sample over and over again. Such a sample is known as a **panel**. Quite apart from avoiding the bother of reselecting the sample, this has the advantage that changes can be assessed more accurately as a result of asking the same people rather than just a similar group. But in practice difficulties arise, and not only the obvious one that panel members may die or move away. When they know that their opinions are to be sought a second time, the way in which these opinions are formed becomes very different, if only because they are thought about more deeply. Or the members may cease to co-operate at all. In the National Child Development Study, the parents of about 92% of the original sample were still co-operating after seven years—a very high proportion in the circumstances. Often the fall-off rate is much greater. Some of these problems are avoided if a proportion of the panel is changed each time so that, as a whole, it does not become 'stale'.

In very carefully controlled conditions, a sample chosen at the discretion of the investigator may be acceptable. Certain areas of a country, for example, may show over a period of time that they contain a representative cross-section of the whole population for a particular purpose, such as producing a percentage swing at an election which accurately reflects the national trend. For many years the State of Maine did this in the United States Presidential elections. But eventually and without warning it ceased to do so. The only safe rule remains: no human discretion at all.

The sampling methods suggested here may be combined together in a variety of ways. One example is given below: the method used for the Family Expenditure Survey, mentioned in Chapter 2. It is a three-stage stratified sample—a fairly complicated one—and it is described here not because every reader needs to understand it in detail but simply to illustrate what is involved in the construction of an actual sampling system on a fairly large scale.

The stratification is based simultaneously on three criteria —region, population density in four grades and an economic indicator derived from the mean rateable value. All the separate administrative areas are classified by this system and divided into 168 strata, each consisting of a group of areas having similar classification. From each stratum one area is selected at a time, with probability proportional to the population of the area, using a random digit system. After being chosen in this way, an area is used four times at three-monthly intervals and is then replaced; 42 areas (that is, a quarter of the total number) are changed each time, so that the pattern of change is the same all through the year. Each set of 168 areas is divided randomly into three groups of 56, which are used in the three separate months of each quarter.

The second stage of the sampling is to choose four units (electoral wards or groups of parishes) from each area, using a fixed interval after a random start, and these four are allocated at random to the four times the area appears in the sample. The third stage is to choose sixteen addresses from the electoral roll for each unit, again using a fixed interval after a random start, with an ingenious modification to ensure that the probability is equal for every address, irrespective of the number of electors at each. If there is more than one family at a particular address, a standard procedure has to be followed in order to make a random choice once again. The system gives the identities of 16×168 families 4 times a year, which is a total of 10 752 families per year.

Finally, a word about the choice of sample size. It has been stressed that this is secondary in importance to the matter of sample quality, but a size has still to be fixed. There is no standard rule: it depends on the time and money available, and on the purpose of the investigation. Anyone familiar with the necessary formulae can calculate the size of sample required to reach the desired degree of accuracy, if the approximate proportions of each group are known. For example, if there are two categories, each accounting for about half the population, and an accuracy of $\pm 1\%$ is required, a sample of 9700 is big enough to give an answer within the stated limits 19 times out of 20. (A brief explanation of the principle behind this statement was given in the last chapter.) This is a fairly high standard, and the opinion polls use samples typically in the range 1000 to 3000, with correspondingly lower accuracy. The size of the population itself does not make very much difference—a sample of 1000 drawn from 10 million is only a slightly poorer guide than if it had been drawn from only 100 000. (In terms of the standard error, the actual difference is about 1 %.)

But, if a relatively small category is to be accurately represented in the sample, a much larger size is necessary. If the group represents about one-twentieth of the population, a sample of about 30 000 is required in order to give a relative error of less than 5 % of the true proportion.

Non-response

Once the sample has been chosen, the next problem is that, except in very simple cases, not every member of it will be able or willing to play his part, whether it be in answering questions, in filling up a form, or in submitting himself to some form of test. He may have died or moved away, as mentioned earlier. He may be difficult to contact because he works a night shift or because he is on holiday. He may have some disability, such as deafness or no knowledge of English. Or he may refuse to answer. The latter group can be reduced to a minimum by legal compulsion, but this is not often possible and is of doubtful benefit anyhow, because the result may be to substitute false answers for no answer at all. Typically, refusals may be of the order of 10 % of the total, but there are many more if the process takes up too much time or if the form or content of the questions is in some way offensive. If the enquiries are made through the post, with no legal compulsion to back them up, the non-response is very much higher—so much so that the method is of very limited value in spite of its cheapness.

The other major source of error is the sample members

who are not at home when the interviewer calls. Second and third visits are expensive, and may not be possible within the time available. The temptation in such cases is to call at the house next door instead, on the grounds that it is just as 'random' a choice as the first one. But to do so is to introduce bias, because the proportion of households with working mothers will be underestimated.

There is no perfect answer to the problem of the not-at-homes, but one method gets as near to it as possible. This is to follow up a small group of them thoroughly, and to use this group itself as a sample of the non-respondents. Any bias shown by the group can then be allowed for in the calculations. Suppose, for example, that a sample of 1000 revealed 420 Conservative, 380 Labour and 200 who could not be contacted. A 10% sample of the 200 might then be followed up, preferably after being chosen randomly, though in practice an area method would be more likely; suppose these showed 6 Conservative, 11 Labour and 3 still not available. The 200 non-respondents would then be divided in the proportion 6 to 11, giving 71 Conservative and 129 Labour. The final figures are therefore about 49% for the Conservatives and 51% to Labour. Non-response in a postal survey may also be followed up in a similar way, by repeated letters, by visits or by a combination of the two.

In Government surveys for industrial and commercial statistics, non-response is still a problem even though it may be due to inactivity rather than unwillingness. The usual practice is to send gentle reminders by letter, telephone and even personal visit, rather than invoke legal powers straight away. Once again, a response may be achieved by force of law, but a reply giving the whole truth and nothing but the truth depends also on goodwill.

In general, the effect of non-response is to create a bias towards the ordinary and conventional. Vagrants, shift-workers, those who spend long holidays abroad, those

whose native tongue is not English, the aged and the chronic-
ally ill—all these are likely to find themselves under-
represented. If the percentage non-response is too great, the
results of the survey may be open to serious question: a
sample of 10 000 of whom only 7000 are successfully contacted is not the same as a random sample of 7000. If the
reason for non-response is related in any way to the purpose
of the survey, the conclusions may be meaningless and, even
if there is in fact no such relationship, a suggestion that there
is one is very difficult to refute. A genuinely random sample
of 700 rather than 7000, but with 99 % response, would be
far better.

The next stage in this brief study of a highly complicated
subject must be to look at some of the faulty methods used.

Unreliable samples

Of the whole long catalogue of faulty statistical methods,
none is more widespread and none is more insidious than the
argument from a non-representative sample.

One example above all others has served as a warning to
statisticians on this point. In 1936 an American magazine
called the *Literary Digest* undertook a colossal survey in
order to predict the result of the Presidential election of that
year. Over 10 million people were asked, but there were two
major sources of error: one was that the response rate was
very low because a postal method was used, and the other
was that the sample was made up partly from telephone
directories and partly from the magazine's own list of subscribers. A clear win for Landon was forecast. But Roosevelt,
many of whose supporters were in the low income groups
which the sampling system grossly under-represented, won
easily. The survey underestimated Roosevelt's share of the
vote by a massive 19 %.

So the statisticians learnt their lesson. But the man in the street seems not to have done so: for him, the consequence of knowing that some statistical methods are open to misuse is that he has developed a distrust of the whole art of statistical reasoning. That he has done so is understandable, but unfair. What he must do instead is to develop a degree of skill in discriminating between the good sample and the bad, and so between the true conclusions and the fallacious ones. One of the difficulties of the situation is that statistical methods are not only used by professional statisticians, and so the reader will often see samples used which are as biased as the *Literary Digest* one. (And also less open to public exposure, as this one was in the election itself.)

The most obvious kind of non-representative sample is the one which is too small, the most extreme case of this being when it consists of one item only. If Mrs P.Q. of Birmingham took Kurakold all through last winter, and did not catch a cold, then it might indeed happen that if you did the same this winter you would not catch one either. But the evidence may be accounted for in a variety of other ways, and the verdict must be 'not proven'.

Consumer research studies are sometimes open to criticism on the same point, though not to the same extent. If a £5000 car is being tested, one cannot use a large sample on grounds of cost, even allowing for the value of the car after testing. And if the one car bought happens to need a new gearbox during the test, the prospective customer will tend to draw the obvious conclusion, which is not necessarily fair to the manufacturer. A sample of 200 cars, with one gearbox failure, would present a very different picture. Of course, the researchers are aware of this, and often use information supplied by subscribers to supplement their own tests, but those who volunteer to help in this way form another non-representative sample.

It is not easy to give a general statement on what size of sample is large enough because, as already mentioned, it depends on several different factors. But to give a very rough rule-of-thumb, in the case where a simple proportion alone is to be estimated, a sample of over 1000 is worth investigating while one under 100 is unlikely to be much use. But, in other kinds of statistical enquiry, a much smaller sample will give useful results: a paired-sample test using two groups of ten individuals, for example. The only safe rule is to consult someone familiar with the calculations involved whenever there is any doubt. Often, however, the inadequacy of the size will be obvious to anyone who thinks about it at all.

When an important change in Government policy is announced, press and television reporters often try to gauge popular reaction by interviewing a small number of individuals, usually less than ten. This may be good journalism but it is not good statistics. The sample is too small, even if it is a random one which is most unlikely; there is also the possibility that bias may be introduced at the editing stage.

If the size is not stated at all, one must judge on circumstantial evidence, particularly the degree of independence and the professional expertise of the body responsible for the figures. An advertiser who says '4 out of 5 can't tell the difference . . .' without further evidence is showing little of either.

The next step is to look for sources of bias in the sampling method. A common type of sample is one which selects itself, such as the people who choose to sign a petition. If 100 000 people sign a petition in favour of capital punishment, it is impossible to make a quantitative estimate of the balance of public opinion as a result—even supposing that the signatures are genuine and all different, which is difficult to be sure about in such cases. This is not to say that petitions are valueless—they are good publicity, to say the least. But if they are to be useful statistically, only people in a restricted (and preferably listed) category should be allowed to sign. Thus a petition from union to management, signed by 90% of the paid-up membership, would be clear evidence; additional signatures of union members in a different works would add little.

Other examples of self-selected samples are letters to newspapers and Members of Parliament for and against a topic of the day, smokers and non-smokers, members of a union or professional association when membership is voluntary, and volunteers for whatever purpose. Animals, birds and fish that get caught are self-selected to some degree, by being either slower or more inquisitive than their peers. The self-selection may be such as to cause no bias in the results, but the possibility of bias is always there.

Those who vote at an election form a self-selected sample of the electorate—still a sample, even though it may include between 70% and 80% of the total. Thus even a perfect pre-election poll of voting intentions could not predict the actual voting figures—at least unless the 'don't knows' in the poll were just those people who abstained in the election, which is most unlikely.

A sample consisting of the people known personally to an individual is non-representative—a fact which is obvious enough, though it does not seem to deter people from making an endless variety of judgements on their fellow human beings on the basis of opinions formed about such a sample. Sometimes this is done without any attempt at balance: the Royal College of Physicians report on Smoking and Health commented that smokers often reject warnings of the risks they take, saying that they know of heavy smokers who have remained well to a ripe old age. So no doubt they do, but some difficulties should be pointed out before conclusions are drawn. The sample is too small, it is self-selected, no one who has died of lung cancer could by definition be a member of the sample at the time of the enquiry, and it is the *proportion* of survivors among the heavy smokers that counts, not just the fact that there are some.

Another common non-representative sample is that of press reports. If, for instance, a wild animal kept as a pet makes a violent attack on a child, then of course it makes news and so public interest in the subject is aroused. So, in turn, any subsequent incidents involving dangerous pets begin to attract attention when, but for the one exceptionally serious case, they would not have been considered newsworthy. Readers of these reports may come to the false conclusion that there has been a sudden increase of such incidents.

Similarly, an item may reach the news pages only as a side effect of a debate on a different matter. When the United Kingdom decided to test the possibility of staying in the West European time zone all the year round, it was predicted by opponents of the change that accidents to children would increase because some of them would have to walk to school in the dark. So the accidents which did occur were reported in national newspapers when in ordinary circumstances they would not have been, and no valid conclusions could be drawn without further information.

A representative sample from a non-representative sampling frame is quite common, since many of the available frames are, as already mentioned, far from complete. A newspaper report claimed that 9 out of 10 schoolmasters were opposed to the raising of the school-leaving age on the date planned—but the 200 teachers in the sample were selected from the membership of only one of the four unions in the profession. The true proportion might therefore have been only 2 or 3 out of 10, not 9.

A low response rate, for which the possible causes were mentioned earlier, means that the sample actually used is non-representative, however good the sample design. Even if the rate achieved is 99%—an exceptionally high figure—

the effect may be to cause an error much greater than 1%, because the ones omitted will have been left out as a result of having been in some way non-representative. A more usual response rate is in the range 75% to 90%, and any results derived from samples with less than 75% response should be treated with great caution. As with some of the other figures quoted in this chapter, no general rule like this can be made up to suit all possible situations, but 75% is a reasonable guide. One of the many defects of quota sampling, incidentally, is that normally no record of the response rate is kept.

To sum up: the onus of proving that a sample is representative falls on the statistician, and the only way in which he can give such a proof is by using a randomly chosen sample, drawn from an accurate frame, in its entirety. In practice this can never be done, and the methods both of minimising the error and of assessing its extent require great skill. The sources of error are many and various, but the most important is the introduction of the human factor instead of a random choice.

When the sample has been chosen, the next step may either be to take measurements or to ask questions. Measurement has been discussed earlier; the matter of asking questions is the subject of the following section.

Questionnaires

The purpose of a questionnaire is, obviously, to transfer information from the mind of the respondent, or from his written records, to the investigator, in such a way that it can be analysed in some numerical form, either by counting or otherwise. Equally obviously, the quality of the result depends on the accuracy of the transfer. But, even in the simplest matter, the process is not as straightforward as it may appear.

The first point is that the questions must not be loaded in

any way. It is not necessary for the originator of the survey to be without bias himself, though it helps. (A newspaper of known political leanings will usually commission an independent research body to carry out its opinion polls, partly for this reason.) If the results of the survey are published without a copy of the exact questions asked, doubts are raised at once: the questions must be known to be fair, and the onus of proof is on the writer.

Other requirements are that the questions should use no technical or difficult words and require no calculations; they should be short, unambiguous and designed to elicit precise answers. When asked of organisations rather than individuals, they must be compatible as far as possible with the answers which the organisation's routine records are capable of giving—it is no good asking for information which is simply not available, or which would take so much trouble to find out that the essential goodwill would be lost.

Often these requirements (particularly brevity in relation to the others) lead in different directions, and great experience is needed in order to choose the best wording. Almost any question addressed to a child is loaded to a small extent, simply for the reason that it is asked by an adult. Also, problems arise if a parent is present or is allowed to see the questions; for different reasons, problems also arise if the parent is *not* allowed to see the questions. There are problems of other kinds in asking questions of very old people: it is more difficult to explain to them the purpose of a survey, and to convince them that it is confidential.

Both the type and the order of the questions are important. We can see why by looking at the different ways of getting wrong answers. In the simplest type of factual question, such as 'At what age did you leave school?', one would like to say that only true answers would be given. But the leading British authority in this field, Professor Sir Claus Moser, comments that the results would in practice be a mixture of 'fact, wishful thinking, vague recollection, and a desire to give the answer the interviewer is believed to be looking for'. Note that this is for simple, factual, non-emotive questions—and very few questionnaires consist solely of these.

Those who do this kind of work very soon find that people do not always tell the truth. If the questions are too personal, false answers are given. If the respondent is not wholly convinced that his answers are confidential, he may give an answer which conceals the true facts. If he begins to resent the questions for any reason, he may give short and therefore misleading answers, or false ones, simply to bring the interview to an end as soon as possible without appearing to be rude.

To illustrate the problem of a factual question which has personal connotations, consider the example of a survey of magazine readership. The results from such a survey can be compared with the known circulation figures, though exact agreement is not to be expected because some copies are read by several people. But in some cases the two figures are quite incompatible—people just will not admit to reading *Playboy*, though they go out and buy it in their thousands. It would assist in the interpretation of such surveys if the results were published in the form '60% *say that* they read the Daily Blurb', rather than '60% read the Daily Blurb', but it does not often happen.

There is an obvious danger of bias in questions about income. Surveys of family income and expenditure have shown that income tends to be understated by about 10%, and expenditure to be overstated by about 5%; the latter figure would be rather greater but for the fact that overestimation of amounts spent on essentials is partly offset by understating expenditure on non-essentials such as alcohol and tobacco.

Hypothetical questions ('What would you do if . . .?') are clearly open to error, though one example of it is well known—that of poll questions of the form 'If there were a general election tomorrow ...'. Another example, the Family Intentions Survey, was mentioned in Chapter 2.

Opinion questions are more difficult still, both in choosing the right wording and in the evaluation of the results. The answers depend, more than in factual questions, on the sex, colour, age and general attitude of the interviewer, and they are more sensitive to slight changes in wording. Because of the effect of the interviewer's attitude, it is essential that precise instructions should be given about the procedure to be followed. It is necessary to define, for instance, the degree of prompting which is allowed, and the action to be taken when a respondent asks for a question to be explained, or wants to give a qualified answer. If such matters are left to the interviewer's discretion, it can have a considerable effect on the proportion of 'don't knows', and smaller effects in other ways. The whole matter of the interviewer-respondent relationship has a major bearing on the quality of the results, and it has itself been the subject of statistical investigation.

There is a useful technical distinction between 'opinions' and 'attitudes', in which opinions are noted simply as agree/ disagree/don't know and attitudes are recorded in a way that distinguishes between strong and less strong beliefs. For example, 'Do you think that hanging should be a punishment for murder?' is an opinion question, and the corresponding attitude question would be 'Which of the following best describes your reaction to the suggestion that hanging should be a punishment for murder: strongly agree/agree/ neither agree nor disagree/disagree/strongly disagree/don't know?' The attitude question gives a much clearer guide in the frequent cases where it is helpful to be able to assess the intensity of public feeling on a subject, rather than just the balance of numbers.

A possible source of error in questions having a numerical answer is unintentional rounding off, particularly where ages are concerned. To a child, the day when he reaches 7 rather than 6 is important, but an 87th birthday may not be of such consequence and rounding to 85 or 90, or even 80, may be the result. In a large sample, the extent of the rounding may be estimated by finding the proportion of all the ages which end with 0 or 5, but this information may not be much help in trying to put matters right. A bias towards even numbers also occurs, but is rather smaller in extent. One way of trying to avoid the difficulty is to ask for the year of birth; the danger here is that a few even larger discrepancies (such as a 10-year error) may pass unnoticed.

Further complications still remain. For ease of interpretation of the results the answers must usually be specific, but respondents often want to qualify one of the 'model' answers, for instance by saying 'Yes, but only for the murder of policemen and prison officers'. Sometimes it is possible to allow for this in the questionnaire, but there is a limit to the number of variations because of the need to be brief. This limitation is one of the strongest arguments against the widespread use of the referendum as an instrument of government: the kind of questions which can be answered by selecting one out of a short list of precisely defined alternatives, without omissions or overlaps, and with no need for qualified or conditional answers, are not usually the questions which governments are responsible for answering. On any finely graded and many-sided matter, such as the use of animals in medical research, satisfactory questions are impossible to construct.

The problems caused by having a large proportion of 'don't knows' have already been mentioned. Careful wording and ordering of the questions helps to keep numbers to a minimum: very few people refuse on principle to answer any question at all, and in fact many feel flattered to be asked. But too many questions, or intrusive ones, tend to produce a lower response rate. If the questionnaire is to include some questions of a personal nature, therefore, they should be kept till the end; this allows time for the interviewer to gain the respondent's confidence, and means that some information will be gained even if the personal questions are unanswered. There is less scope for such niceties in a postal questionnaire.

To sum up:

(*a*) the report of the survey should include the actual questions;

(*b*) the questions should show no evidence of bias;

(*c*) the questions should be simple enough to answer without too many reservations;

(*d*) there should be no reason for bias in the answers.

As with the earlier summaries of the requirements of sampling frames and sample designs, this represents an ideal which any actual questionnaire must fall short of, if only because human nature itself is a reason for bias in the answer to almost any question.

The estimation of those errors which are unavoidable whenever sampling is used has already been discussed. To estimate the extent of the avoidable ones is very difficult, and requires subjective judgement, based on experience combined with just a little circumstantial evidence. Sometimes it is possible to check the answers in individual questionnaires for self-consistency: a family's income and its expenditure, for example, if saving and borrowing are also allowed for. Alternatively, aggregated figures may be open to independent check: numbers of cigarettes, for example, reported to be smoked by a sample of the population (children included, preferably) can be scaled up and compared with the national total which is known fairly accurately from excise duty figures. A more elementary example is that exports from country A to country B should tally with B's figures for imports from A. It is a salutary lesson, both for statisticians and for students of human nature, to discover how great are the discrepancies which such checks often reveal. The under-reporting of expenditure on alcohol in the Family Expenditure Survey, mentioned earlier, is fairly consistent at about the 50% level.

Pilot surveys

When a statistician meets a new problem, or an old problem with a new feature to it, he can rarely guarantee to get everything right first time, and so a preliminary try-out is useful. Typically, this will take the form of a small-scale version of the final scheme, known as a pilot survey. It may include a test of the sample design, or may be simply a test of the questionnaire using any convenient group out of the population later to be sampled.

The kind of defects which may show themselves are those which have been mentioned already; two in particular are the questions whose wording is not sufficiently clear, and the questions for which respondents wish to give qualified answers. Also, questions which reveal an adverse 'consumer reaction' can be detected when they can be reworded, repositioned or omitted altogether, so that the non-response proportion can be kept to an acceptable level.

Just as a sample survey has advantages over a census, for reasons discussed in Chapter 2, so a pilot survey has advantages over a larger one. It can be carried out by more experienced interviewers, who can detect errors which others might not notice. Detailed checks can be carried out more easily. And open-ended questions can be used: the answers given to these can be used to suggest the list of alternative answers to be included in the final version, in which open-ended questions cannot normally be used on account of the difficulty of coding the answers for computer input.

For the Census of Population, extensive pilot surveys are

carried out. In 1971, however, many things went wrong which the preliminary work ought to have given warnings about. What seems to have happened is this: the reaction of a small number of individuals, when invited to help in a survey carried out without much publicity, whose purpose was carefully explained by an experienced interviewer, was favourable; the collective reaction (influenced by the mass media and the actions of a few leading citizens) of a large number of people to a full-scale, highly publicised census carried out by specially recruited and underpaid enumerators, with the threat of legal action if they did not co-operate, was quite different. Statisticians, like everyone else, learn by experience as well as from books, and in this case the experience was rather painful.

8 Statistical experiments

In most of his work a statistician is primarily an observer, searching out information which already exists either physically or in people's minds. He hopes and expects that the information he is able to write down, analyse and explain to his readers will influence the future course of events, but only rarely is he able to control matters himself. The statistical experiment gives him the opportunity, and when conditions are right it can be a very powerful instrument.

Essentially, the problem it solves is one of comparisons. A new teaching method, for instance, is tested by comparing its results with those of the method it is designed to replace or improve. We are not concerned here, of course, with the comparisons in which an objective answer is possible by direct measurement, such as whether one mountain is higher than another, but the kind in which some uncertainty is involved. Other examples of such comparisons are selective and non-selective education, fluoride in drinking water, rearward-facing aircraft seats, hanging as a deterrent against murder, the efficacy of different drugs, different fuels, different fertilizers, different feeding methods for babies and so on. The list is a long one, but not every case is susceptible to investigation by experiment; we start by listing the reasons why this may be so.

First, an experiment may be non-repeatable. It would be of great interest to economists to be able, say, to devalue a currency by 10%, watch what happens, and then go back again and devalue by 20% instead as a comparison. But of course it can't be done (at least not without a computer 'model' of the country's finances, which is now used for such purposes but which is outside our scope here). Even a model cannot give a true picture of a situation which is subject to so many external influences. Similarly, a race or contest of any kind cannot be repeated without altering the conditions, and so cannot be used for an experiment of a genuinely statistical nature. Even if nothing else changes, the players' knowledge of the results of previous contests means that conditions are not precisely the same.

Next, the experiment may be too big and expensive. One could design a statistically satisfactory experiment to test the relative safety of rearward-facing aircraft seats, for instance. This would use an aircraft filled with dummy passengers of whom some faced one way and some the other, which would be guided by remote control into a spectacular crash. To do the job thoroughly, this would have to be repeated at various speeds and angles of impact. It is hardly necessary to explain why the idea is not likely to be tried out.

The third reason concerns experiments which involve human beings. The difficulty here may be either an ethical objection as far as the experimenter is concerned, or an un-willingness to accept the conditions of the experiment as far as the subjects are concerned. (The ethical objections also apply in the case of some animal experiments.) In testing a new drug, for example, the experiment if properly conducted would require the choice of drug to be made on a random basis, with no influence from the doctor and, ideally, even without his knowledge. Except in the case of minor ailments such as the common cold, medical ethics make such conditions unacceptable, and rightly so.

Not only medical experiments are involved. An experiment to compare the social effects of different types of housing, such as high- and low-rise flats, would be of great interest both to the housing departments of local authorities and to sociologists, but in practice people would object to being allocated to one type or the other by a method which was in effect just spinning a coin. Experiments in methods of teaching give rise to particular difficulties in this respect, because neither the children chosen as subjects, nor their parents, may be in a position to raise objections. In general, the only conditions in which a statistical experiment can properly be carried out are those in which the people taking part (and their doctors or teachers, where appropriate) are willing to accept without strong preference either of the things being compared.

Another experiment in which the degree of control would be quite unacceptable is one designed to test the relative effects of heredity and environment on a child's development, a matter which has been the subject of bitter controversy because of its racial implications. One possible method would be to take a thousand or so children born within a few days of each other, and redistribute them to the corresponding group of parents on a random basis; another would be to divide at birth fifty or so pairs of identical twins. But again one must hope that such ideas remain pipedreams.

So to the positive side of the picture—the construction of an experiment which is financially and ethically possible. It is assumed to start with that the subjects are human.

Ideally, the subjects should be chosen randomly from the population to be investigated: smokers, sufferers from rheumatism or those taking a particular educational course. But this is rarely possible, and in practice one has to use a group already in existence, such as the children in a particular school, or a group of volunteers. This is quite satisfactory as long as there is no obvious reason why the group may be non-representative.

The vital point is not the selection of the test group but the way in which it is divided into two parts. One of these is used as the group actually tested, while the other is used as a standard of comparison and is called the control group. This division has to be carried out on a random basis and, when conditions allow, in such a way that the subjects themselves do not know which group they are in. This may seem a rather severe condition, but it is no good telling one group, 'You are to have a new treatment instead of the usual one, to see if you do better than those who still use the old method.' The group's own consciousness of its experimental role will have an effect on the outcome, and the experiment loses its objective nature.

So let us suppose that we have 40 volunteers, all with heavy colds, and we wish to try out a cold cure which consists of 21 tablets for each patient, to be taken at 8-hourly intervals over one week. The volunteers are first numbered in any convenient order, and 40 containers for the tablets are numbered correspondingly. Next, a random digit method is used to select 20 numbers out of the 40. The sets of tablets for these patients are then placed in the appropriate boxes. Into the others are placed sets of placebos—dummy pills, of the same appearance as the others but with no medicinal properties whatever. It is not necessary for more than one person to carry out these preparations and, if he is not involved in the tests themselves, so much the better. The list showing the actual distribution of tablets is not used again until the results are being assessed.

During the experiment itself all the patients are treated exactly alike, and nothing special needs to be said except that it is necessary to record the progress of each patient in some numerical form such as the number of days required for the cold to reach a particular stage of development. Of course, not all the colds will have been at exactly the same stage of development at the start of the tests, but this does not matter because of the random separation into two groups.

When the experiment is over, the figures for the two groups are separated, and a fairly simple test can be used to find whether there is a significant difference between the means of the two groups: that is, a difference which is too big to be accounted for by the routine variations inherent in this type of measurement. All this is completely objective and rigorous, and if the results were satisfactory the makers of the tablets could then put in their advertisements: 'Scientific tests PROVE that . . .' Just for once, such a claim would have some truth in it.

As with the statistical tests described in other chapters, the results are given in terms of a confidence level, say 95% or 99%. For a matter such as this, fairly conclusive evidence is required, and so 99% would be the better choice. The method of calculation will be found in more advanced books.

The procedure just described represents something like the ideal. In most experiments, things do not work out so neatly. For one thing, the nature of the experiment may be such as to make it impossible for the separation into test and control groups to be disguised: a new teaching method, for example, is at once obvious to the people on whom it is tried out. Also, the construction of the experiment and the assessment of the results may not be possible in a way which treats both groups alike: an instructive example of this is in trying to compare methods of teaching reading.

Consider the problem of constructing a test to compare two of these, the look/say and the phonic methods. In the first, the children learn to associate the word forms with pictures which describe them, and in the second the association is between the words and their sounds when spoken. Both methods have been used successfully in schools, but objective comparisons are difficult. Some words are both easy to put into picture form and easy to pronounce, and there is no problem with these. But others such as *window* and *bucket* are easy to draw pictures of while being difficult to pronounce on account of their two syllables, and so are harder for the phonic group. And words such as *gig* and *ban* are easy for the phonic group but difficult to show pictorially. Also, the look/say children will tend to learn nouns earlier than other words, the phonic ones rather less so. The obvious answer is to construct a test with a fair balance, to suit both groups. But this makes the test no longer an objective one, and the proponents of the method which the results indicated as less effective would be able to claim that the choice of words favoured the other side. Whether there was in fact any such bias is not really very important; the essential point is that a *suggestion* of bias cannot be refuted objectively, so that the opportunity to influence people's thinking on the subject is very much reduced.

But most of the defects in statistical experiments derive from the use of pre-existing groups for the tests and for the controls. Here the danger of serious bias is immense and, as in sampling, the use of larger groups is no help in counteracting it. Just occasionally, it is possible to use pre-existing groups in a way which effectively excludes bias, as in the early tests of fluoridation of water supplies, described in Chapter 12. Here the two groups had no detectable differences related in any way to the subject of the experiment, and so could be regarded as possessing all the qualities of a single randomly divided group.

This, however, was a rare case. Similar samples have been used in attempts to compare the merits of selective and non-selective education, by looking at different areas where each of the systems is in operation. But there are many reasons why such attempts have been inconclusive, of which it is sufficient to mention just two. First, the actual differences, whatever they may be, are nothing like so spectacular as the decrease in dental caries after fluoridation—nobody expects changes of the order of 50% or more in any quantitative measure of educational achievement. Secondly, there is no general agreement as to what those measures should be, as examination performance is certainly not the only one.

Cases such as this are the norm, and those like fluoride the exceptions. Two groups of volunteers are no better than two specially chosen groups, at least if they are aware of whether they are volunteering for the test or as a control, since someone becomes a volunteer because of certain qualities and interests which may be related to the test. The volunteering may be done by proxy: certain schools which use out-of-the-ordinary teaching methods manage to produce examination results which are consistently higher than the national average, but this proves only that it is the parents of a non-random group of children who have volunteered to send them to these schools, not that the teaching methods themselves are any better.

This last example is one in which experimental results for one group only are compared with independent figures, derived from Government statistics of examination results. This is sometimes useful, but it requires the test group to be randomly chosen from the same population, which is rarely possible. There is a further difficulty in that the main group and the control group are not tested under the same conditions. In fact, the test conditions for the control group are outside the influence of the tester altogether, and the effect of this may be to circumscribe the conditions for the actual test to an unacceptable degree. And, if the conditions are altered, no comparisons are possible.

If the test and control groups are the same, and the comparison is made over a period of time, all sorts of difficulties arise. The subjects will know which group they are in, for a start; they will have, or develop, personal preferences; and they may have a preference for old and familiar ways as such, or for new and up-to-date ones. This last point is true of experiments in general, of course, and is not only true of conscious preferences. A cigarette manufacturer once asked in an advertisement for a test to be made of the mildness of his cigarettes: one puff of any other brand, followed by one puff of the brand advertised. But if one tries this out with two identical cigarettes, the second always seems milder.

This example suggests, in fact, one way of using a single group for both test and control, and yet avoiding the difficulties mentioned above. This is to divide the group, not necessarily randomly, and do the test with one half while the other half forms the control; when this is done, exchange roles and repeat. If the group which starts with the test does better when it is given the control, while the other group also improves when transferred from the control to the test, no useful conclusions can be drawn. But if both do better on the test than on the control, the result is fairly clear. However, this approach is not always possible, as the experiment may be of a non-repeatable nature.

The effect of the involvement of subjects in the experiment, in the sense that they develop an interest in its outcome, was demonstrated in a remarkable experiment which took place in the suburb of Hawthorne in Chicago, and has become known as the Hawthorne Effect. Tests were carried out in an assembly room to measure the changes in output which occurred as a result of alterations in the lighting. Each time a change was made, the output went up. But the changes were not all in the same direction and, when the intensity was set exactly at its original level, output went up again. The subjects had been so keen to co-operate that the changes in output took place almost independently of the changes in the lighting.

The experimenter who uses animals has fewer problems to deal with—at least his subjects react to the actual test and not to the fact that they are being tested. Ethical considerations of course still exist, but not to the same extent as for humans. The number of such experiments is very large—currently over five million per year in the United Kingdom, according to the official figures which probably underestimate the true ones to some extent.

The reason for this large number is principally, of course, the fact that the ethical difficulties are so much less, and that the deliberate infection of a randomly chosen test group is possible under certain conditions. The development of a cure for any form of cancer, already immensely difficult, would be very much more so without such tests. But a conclusive experiment with animals is not directly transferable to humans, even qualitatively. Substances which are carcinogenic in animals are not necessarily so in humans, and vice versa. Rabbits can eat Deadly Nightshade without harm although it is poisonous to humans, and rats and mice deprived of vitamin C do not develop scurvy. So the effects (including side-effects) of new drugs are found by tests on several different kinds of animal before they are tried out, very cautiously, on humans. The inadequate testing of drugs in a few well-known cases such as Thalidomide has, of course, led to much greater vigilance in these matters.

A point which requires as much care with animals as it does with humans is the division into test and control groups. A random choice out of a group of a few hundred mice is not as easy as it sounds: catching them one by one means that the slow ones get caught more easily, and allowing them out through a small door is no better as the inquisitive ones will emerge first. The only way to be really sure is to number them and use a random digit method, but in practice a systematic method such as putting them alternately into two groups would probably be near enough.

Experiments in agriculture and horticulture remain to be mentioned. The basic idea is the same as before, but the physical arrangement of the test and control groups requires some care. It is usual to arrange that the conditions for both groups—type of soil, watering and drainage, sunlight and warmth, and nutrients of various kinds—are the same except for the one item which is the subject of the test, so that there is no doubt about the reasons for any differences which are observed. But, if this is to be so, it is no use dividing the available area neatly into two, because one end may receive more sun or better drainage. Instead, the area is divided into squares or rectangles, and alternate sections are used for the two groups, as on a chess board. For more than two groups (if it is desired to test several combinations of variables at the same time), an arrangement called a Latin square achieves the same result. The more areas used, the more accurate the results should be, but the experiment becomes correspondingly more fiddly to operate. In certain rather complicated experiments, a mathematical technique called the analysis of variance can be used to advantage.

Finally, one particular experiment which is recorded in the Book of Daniel is worth a mention. Daniel and his three friends were captives of the king of Babylon, and had been chosen for the Royal service. As a result of this, they were fed on rich food which they did not like. Daniel therefore proposed an experiment: he and his friends would be fed on vegetables, with water instead of wine, and after ten days their appearance would be compared with that of the other young men at the court. The king's steward, although concerned for the safety of his head should the experiment fail, was persuaded to agree, and at the end of ten days the four men looked healthier and better nourished than the others. Daniel's confidence, of course, rested in something rather higher than the laws of probability, but all the same he had proved his point. The experimental procedure is open to criticism on a number of counts, and the time allowed was certainly rather short. However, as it took place over 2500 years ago, it must be regarded as well ahead of its time.

9 How to read a statistical argument

This chapter takes the form of a statistical fault-finding test. It is designed to search out the flaws, fallacies and omissions which are commonly found in statistical data and reasoning, and to assist in the interpretation of statistical arguments generally. In the interests of avoiding too many cross-references, some points are mentioned briefly even though they are covered in more depth in other parts of the book; in the case of some major topics, reference is made to other chapters.

A fault-finding test, if it is to be comprehensive, also has to be fairly severe, and the detection of a fault does not necessarily imply that the conclusion should be rejected. The effect is usually to weaken a link in the chain of argument, not to break it and render the whole chain worthless.

The words behind the figures

Who makes the claim? This must be the first question to ask, in order to distinguish between a report written by an independent observer of undoubted professional competence and, say, a statement by an advertiser about his own products. To give an example, if one were to read that a medical research body had discovered a certain substance to be carcinogenic, one would be quite willing to accept the statement as fact. But when an organisation concerned with the manufacture of dairy products suggested that an ingredient of margarine might be carcinogenic, the margarine manufacturers could hardly be expected to accept the statement. In fact, when this happened they were able to show without much difficulty that the claim was based on highly suspect evidence.

It is useful to form, right at the beginning, a general impression of the argument. If it is in a newspaper or magazine, one should ask whether the writer appears to understand it himself or whether he is using his source material uncritically. A headline such as 'Garages demand £3 ransom' (published the day this is being written) suggests the latter, and in fact the article itself showed that the charge, which was for service when the garage was officially closed, was a justifiable one.

Next, are the key words properly defined? A statement that two-thirds of the world's population suffers from malnutrition is almost valueless unless it is known what the writer means by malnutrition, either by giving his own figures or by quoting a standard definition. (The statement itself, by the way, is still frequently made but is now recognised to be based on too high a minimum nutrition level.) Numerous examples of other words which are sensitive to changes of definition will come to the reader's mind, such as director, trade-unionist, engineer, student, television viewer, profit, income, productivity, intelligence, and any word indicating a person's adherence to a religious body or set of beliefs.

Technical terms such as 'essential oils' may mislead the outsider. Another example used in this book is the word population, which is used by statisticians to refer to collections of objects as well as of people. Geographical terms such as land area are often difficult to define—should this, for example, include inland water and, if so, to what limit of size? Loaded descriptions such as 'unwanted babies' at once indicate special pleading, so that any claims about the numbers of them should be treated with caution. Adjectives which, taken literally, imply an absolute standard—such as pure or heatproof—acquire meaning only by common usage sometimes confirmed by legal definition.

Categories

The three questions to ask about categories are whether the grouping is emotive, whether it is subject to personal bias and whether an objective choice is possible in marginal cases. Division into standard blood groups is a straightforward example, with no emotive content, no reason for bias and (in the principal blood group systems) no borderline cases. At the other end of the scale is the classification of people by racial origin, which is impossible to carry out completely even if one believes it to be practically desirable and morally justifiable.

Apparently simple cases such as sexes and 'fatal' accidents can give rise to problems, and the division between married and single is an uncertain point on a scale which includes divorced, divorce pending and legally separated. (Even this assumes that the subjects are telling the truth, and it is well known that a man legally married but separated is more likely than his wife to say that he is single.) Legal definitions may differ from a layman's expectation, such as the classification of a clergyman as self-employed for tax purposes. Popular misconceptions are frequent, such as the belief that rhubarb is a fruit when it is a vegetable; sometimes a correct classification may be unhelpful, such as putting dolphins in the 'mammal' class with the lions, instead of the 'fish' class with the sharks.

Industrial classification poses endless problems, which became important to others as well as statisticians when a selective employment tax was introduced. The tax was payable for workers in service industries but not for those in manufacturing industries and, even though the planners tried hard to produce a foolproof definition of 'manufacturing', the borderline disputes were numerous. For the routine recording of industrial statistics, there is a system called the Standard Industrial Classification, but many firms do not fit neatly into a single category; for the purpose of the statistics of employment, therefore, the classification of the head-quarters staff is not easy.

Classification by occupation or status is equally difficult, and the results are only useful in the most general terms. In Britain there are two widely used systems of what are called socio-economic groups: those of the Registrar General (used in Government surveys) and those of the Institute of Practitioners in Advertising (used in market research). The first of these can be used either in the complete version, which has 223 categories, or in a simplified one. The status of wives is normally determined by the occupations of their husbands, though how long this convention will survive in the days of feminine liberation is uncertain.

Other organisations have their own methods of classification. The Open University, for example, has a self-coded system on its application forms so that the background of its students can be assessed statistically. The possible anomalies are obvious enough not to need spelling out in detail, and one example will suffice. The Open University class labelled Communications and Transport includes both airline pilots and office messengers, a group too diverse to be useful without subdivision. In the days of king, barons and serfs, matters were much easier.

Residual classifications are particularly prone to error. If there are five divisions, and the items in group E are those left over when A, B, C and D have been filled, then the error in E will be caused partly by the combined errors of the other four and partly by the omission of any further categories. A common example is in religious divisions: if 40% of a country's population are thought (probably inaccurately) to be members of the Roman Catholic Church, then the other 60% are likely to find themselves classified as Protestants irrespective of any church allegiance they may have.

Category definitions may vary, both from place to place and (either legally or by usage) over a period of time. Statistics of different types of school, for example, are difficult to interpret because, in some areas, schools are called comprehensive even if they do not have an all-ability intake. Criminal statistics are particularly affected by changes in definition: raising the age of criminal responsibility may give a false impression that juvenile crime has decreased, and redefinitions of murder have contributed to much specious argument about the deterrent effect of hanging.

Unless the grouping is entirely unambiguous, one must also ask who has done the classification. Even the most straightforward system can fail: an interviewer who only needs to find a male old age pensioner in order to complete a quota sample before she goes home to tea may be tempted to use someone else instead. As a general rule, those who do this work should be trained, experienced, disinterested, as few as possible in number and paid by time, not results. (The failure to follow the last point in the 1971 Census of Population led to many errors and much ill-feeling.) Those with personal bias and those who record the information only as an incidental (and probably rather tiresome) part of their jobs produce much less accurate results.

Sometimes attempts are made to classify indirectly when the subject is a sensitive one. It is clearly desirable, for instance, that children whose language in their own homes is not English should receive special attention in school and therefore, for planning purposes, one must find out how many of them there are. But one could not reasonably find this out by observation or direct questioning, and so an indirect method must be used, such as finding out the countries of birth either of the children or of the parents. But, even if this information itself was found with reasonable accuracy, which would be far from easy, the problem would not be solved because the relationship is not an exact one.

Units

A unit is more easily defined than a category, but care is still required. The obvious first requirement of a good unit is that it should be the same in all circumstances. Although the day is long past when the cubit was defined as the length of a man's arm from elbow to finger, some anomalies remain, such as the fact that the United States gallon is about 17% less than the Imperial gallon. The change to metric units should bring such muddles to an end. One that may continue, however, is that a U.S. billion equals a thousand million while the British billion is officially a million million. Although the American unit is a much more practical one and is widely used in Britain, there is as yet no move to recognise it officially.

The precise standardisation and ease of reproduction of a unit are clearly important, and the history of weights and measures is largely an account of their development. International standards for the metre and the kilogramme have been kept in Paris, with official copies elsewhere, but the metre is now defined in terms of the wavelength of a certain type of radiation. Similarly, the unit of time is now defined in terms of the period of another type of radiation. Length and time can therefore be measured in a laboratory, to accuracies of the order of one part in a thousand million, without reference to outside standards.

The units of mass, length and time lead directly to units for velocity, pressure, energy and so on. If a quantity is measured in different ways—say, fluid flow in litres per minute or cubic metres per day—a simple conversion factor relates the two. Other units such as those for temperature, luminosity and electrical measurements can be defined with precision. It is in less scientific matters that problems occur.

The simplest form of measurement is by counting, which

sounds simple enough but may be misleading unless the items counted are all the same. Oranges may be easier to sell by number than by weight, but it becomes impracticable when they are of widely differing sizes. A typical automatic traffic counter records the number of axles which pass over it, so that two bicycles count more than a three-axle lorry. Less obviously, counted figures for house-building or vehicle production may be easy to read but are of limited value, particularly when making fine distinctions, because large and small items all count the same. A unit, in other words, should be homogeneous.

Thus it may be useful for some purposes to group together murder, malicious wounding and similar offences in a single category called crimes of violence, but a clear trend in one particular type of violent crime may be masked by so doing. Audience research tends to equate the casual listener with the enthusiast, so that the proportions of each have to be estimated in other ways in order to form a true description of the audience. Most voting methods are open to the same objection, but here the drawback (if it is one) usually has to be accepted. Attempts have been made to give varying degrees of voting rights to different groups, but they have not proved popular.

Another example is that one of the measures used as an indicator of industrial unrest is a count of the number of strikes which start over a given period of time. The current annual figure for the United Kingdom is about 2500, but it could be reduced by the simple expedient of holding a single year-long general strike. In such an extreme case the misleading nature of the measure would be obvious; the bias in more normal conditions is perhaps more important just because it might escape notice.

Even the use of a recognised unit is no guarantee that it is being used in a precise and unambiguous way. It is conventional, for instance, to compare traffic densities in different countries by quoting the number of vehicles per kilometre of road. But, in making a true comparison, the precise definitions of both 'vehicle' and 'road' are critical, and they vary from one country to another.

Some units of time are not homogeneous, a month being the obvious example. As well as the effect of the different lengths of the months, the number of *working* days may be more relevant than the total number: thus a 20% improvement in output from April to May might be due simply to the fact that April had five weekends including the Easter one while May had four weekends only. The effect of the extra day in a leap year is small. But some figures are recorded for a whole number of weeks, so that a fifty-three-week 'year' sometimes has to be allowed for in order to catch up on the lost days.

Other units may also vary. Measures of volume correspond to different quantities by weight on account of the expansion caused by pressure changes (particularly with gases) and temperature changes. Solids in powder and granule form also change in volume due to shaking in transit. Some materials (notably wool) absorb water from the atmosphere, thus increasing their measured weight.

Sometimes more than one unit is required for a full description. Dietary standards, for instance, are sometimes expressed only in calories, but the adequacy of a diet is determined also by the quantities of protein, minerals and vitamins.

Certain industries use their own units, which may be clear to those who use them from day to day but are often far from clear to the outsider. A barrel, for example, means a different quantity depending on whether it is a unit of measurement for beer or for oil. There may have been good reasons for using such units when barrels were actually used, but in the era of bulk transport they simply cause confusion. One hopes that terms like these will soon fade into history, following the examples of some which were used in the paper and pharmaceutical industries until recently.

Finally, different measures may be used for the same thing, with no fixed conversion factor. The reason may be convenience of measurement or simply to suit a particular point of view—to make the figures look more (or less) alarming than they actually are, or to make a comparison appear more favourable to one's own side. Imports and exports may sometimes be measured most conveniently by weight, by volume or by value. Convenience is not, perhaps, the criterion in the matter of aircraft accident statistics: here the number of accidents, fatal accidents or fatalities may be compared with the total distance covered, the total flying time, the number of stages or the number of journeys (for each passenger in all cases), in any combination. Since any one of these twelve choices is open to assorted variations of definition, the likelihood of achieving international standardisation is remote.

To sum up, a unit should be precisely defined, homogeneous, comprehensible to the layman and, as far as possible, comparable with other units designed to measure the same thing.

The figures themselves

The first question to ask at this point is whether the figures are up to date. Two separate times are involved, those of collection and publication. The publication date can usually be checked without difficulty, but it may not be so easy to know the date (or range of dates) of collection. In routine statistics there is often a regular pattern: the number of those registered as unemployed in the United Kingdom, for instance, is published on the Thursday after the third Monday of each month, but the count is actually made on the Monday of the previous week. It may be helpful to know whether such a figure is worked out on the same day of the week or on the same date in each month, in case the number is usually higher on some weekdays than others.

Sometimes the collection of the data may take place on a single day, and sometimes as a running total. In meteorological work, for example, barometric pressures are recorded as instantaneous values, while rainfall is measured as a total over a given period of time. In some cases, either method may be used depending on the purpose of the investigation, such as the flow of a river which may be recorded as an instantaneous rate or as an annual total quantity.

When totals are stated, the limits of the period must be stated also: for instance, whether the year is a calendar or a financial one. Often such figures are published for both monthly and annual periods, and for the most up-to-date figures the monthly totals may be better. But care is necessary on account of possible seasonal differences.

Sometimes there is a long delay before publication, because of the complexity of collation or the failure of those responsible to supply information on time. When this is so, it is a common practice to publish preliminary figures fairly quickly and corrected ones at a later date. One interesting example is that a 'fatal' road accident is usually taken to mean that death occurs within one month, so that daily or weekly returns could not be guaranteed correct.

As far as dates are concerned, official publications are rarely at fault. It is when their figures are quoted by others that the trouble starts. A general reference book, for instance, may quote a population figure from the United Nations *Demographic Yearbook* without stating the date of the estimate as given in the *Yearbook*.

Next, how were the original figures calculated—by counting, measuring, estimation or guesswork? Crowd figures, for example, may be found by a variety of methods giving results varying between near-total accuracy to estimates five or more times too large. Press estimates of the attendance at one political rally were in the range 40 000 to 200 000; aerial photographs later showed that the number was in the range 23 000 to 25 000.

Counting ought to be reasonably accurate, but human errors do occur, as anyone who (as an umpire in a game of cricket) has tried to count six balls in an over will know. When a recount is necessary in a closely fought election, further counts rarely give the same figures.

A possible ambiguity is the distinction between counting a number of intervals and counting the end-points of the same intervals. Thus it takes a clock twice as long to strike eleven as it takes to strike six, because the time is measured by counting the intervals between chimes rather than the chimes themselves.

This curiosity also appears when orderings are used, as for instance in counting (inclusively) the number of days from the tenth day of the month to the twentieth, which is eleven and not ten. Conventions in counting differ from place to place: for example, the usual way of stating the Christian doctrine of the Resurrection is to say that it took place on the third day after the crucifixion, following the Jewish system of counting, while the more familiar usage would be to refer to Sunday as the *second* day after Friday. Similarly, some countries use the 'Chinese' method of measuring ages, in which a new-born baby is credited with age 1, and the figure increases by one every Chinese New Year rather than on the anniversary of his birth.

The problems of measurement open up the whole subject of accuracy, which is covered in more detail in Chapter 3. The important point to note here is that the degree of accuracy has somehow to be estimated, as reliably as circumstances allow, if the figures are to be put to the best possible use. All too often, no estimate is given by the person who is in the best position to do so—the one who takes the measurement or, in the case of a team operation, the leader of the team. A good example of the way things ought to be done is the crowd count mentioned above. An estimate from aerial photographs would certainly give problems of interpretation, for several reasons, but the extent of the possible errors could only be judged by someone who had examined the photographs with some care. A total in the form of a range of possible values, such as 23 000 to 25 000, was in this case the best form of estimate. When no indication of accuracy is given, the reader must judge for himself, usually on rather limited evidence.

Why were the figures collected at all? This may give an indication of any tendency for errors to be more in one direction than the other. The income figures published by the tax authorities, for instance, may safely be said to include far more underestimates than overestimates.

Who did the measurement? Just as for classification into categories, there is always some scope for personal judgement, and so one should ask if the work was done by someone trained and experienced, aware of the precise limits of what was to be included and instructed about what to do in doubtful cases. Also, did they believe that it mattered whether their work was conscientiously done? Was it a major part of their jobs or only an incidental one? Were any checks carried out? One does not always know the answers, but their influence on the accuracy of the results is great.

Have the figures been rounded? It is usually possible to tell from the number of zeros at the ends but, if rounded figures have been added to unrounded (or differently rounded) ones, the appearance may be misleading. Rounding may also be hidden by a change of units: if a news agency reports a serious fire in New York causing damage estimated at $10 million, the figure may appear in a British newspaper as, say, £4 396 000. The use of an exact exchange rate gives a misleading impression of accuracy.

Have the figures been adjusted in any way? This is usually done only by professional statisticians who understand the process well, and so does not normally cause any trouble. But it is as well to know when it has been done. Briefly, the idea is to alter the figures in such a way as to remove any routine and predictable changes, so that the presence of an unpredicted change may be seen more clearly. Monthly output, for instance, is often adjusted to allow for the differing lengths of months. A more difficult example is in unemployment statistics, for which a possible adjustment is to cut out the routine seasonal trends, using the figures for earlier years as a guide, so that the underlying trend is visible at once.

A distinction between primary and secondary data is sometimes useful. The figures actually recorded, and direct calculations on those figures, form the primary data; indirect calculations yield secondary data, and such figures are in general terms correspondingly less reliable. The boundary line is an uncertain one, but an illustration may help to make the distinction reasonably clear. A census total is an example of primary data, and likewise the totals for births, deaths, immigration and emigration for the twelve months beginning on census day. From these five figures, an estimate could be made of the population at the end of the twelve-month period, and this would be considered as secondary data.

Has there been any interpolation or extrapolation (that is, any estimates of values either between the ones actually measured or beyond the range of them)? Direct interpolation, using simple proportion, usually gives acceptable accuracy unless the series of measurements is very irregular; for an exponential series, a more advanced method called logarithmic interpolation should strictly be used, but it is rarely essential.

Extrapolation is different, and should be looked at suspiciously every time it is observed. It is an important technique because it is essential in major planning decisions. The particular problem of the extrapolation of population figures is discussed in Chapter 16. Other examples are the extrapolation of current trends to estimate the future demands for water and electricity, which has to be done for some years ahead because reservoirs and power stations take several years to construct.

A case in which planning by extrapolation proved to be inaccurate is in long-distance air travel, where a rapid rate of expansion during most of the 1960s failed to continue after about 1968. This left the airlines with much surplus capacity, and manufacturers with few of the expected orders for 'jumbo' jets.

The layman has, of course, insufficient information to reach an informed opinion on such matters in all but a few cases. But examples like this should encourage him (in the unlikely event that encouragement is needed) to treat long-term estimates with suspicion. This is particularly so if there is some finite limit to the changes which are taking place, as is the case in fact with the number of airline passengers. So, when a new vaccine is developed for a particular disease and the number of cases reported shows a downward trend, the use of straight-line extrapolation would be wrong because the number could not possibly fall below zero.

One of the key extrapolation problems during the next two decades will be the use of natural resources, particularly metals and fossil fuels. As the exponential graphs of the amounts being used up climb towards the unknown but finite quantities still remaining on earth, the development of alternative materials and power sources will require planning skill of a very high order.

What statistical measures have been used? They are not always used correctly, and if in doubt see the sections on each of them in other chapters. If a sampling method has been used, many questions have to be satisfactorily answered before the conclusions are accepted at face value; if in doubt on this one, Chapter 7 should give at least some of the answers.

Finally, it is instructive to find out whether the figures are self-consistent. If a total is given, is it correct? Do percentages add up to 100? Errors do occur, though more often in transmission than in an original. But occasionally a basic error may be spotted. On the label of a well-known packet soup, for instance, it is stated in large type that it cooks in five minutes—but closer inspection shows that cold water must be mixed with the powder and brought to the boil before the five minutes starts. The total time could not be less than about eight minutes. However, only the trivial errors like this one are revealed just by looking at the published reports; the more important ones require a rather deeper understanding of the nature of statistical reasoning.

Costs

For most people, the idea of a cost is not a difficult one when it concerns items of personal expenditure. But this familiarity can lead to overconfident acceptance of figures quoted by others for bigger items. A full treatment of economic statistics is beyond the scope of this book, but misuse and misunderstanding of the concept of a cost is so widespread that some aspects of the subject must be mentioned.

Is the stated cost an actual book-keeping figure or an estimate? This, at least, is a distinction familiar to most people, although the figures which appear in company accounts are often less real than they seem, for reasons too technical to go into here. When looking at an estimate, one must ask first who made it and for what purpose; this applies to estimates both of past and of future costs. For an estimate is not a quantity which can be calculated objectively, and the consequences of personal bias may be considerable.

The most startling examples of this can be seen in the figures for Government projects such as *Concorde* and the Channel Tunnel. In matters such as these, the first estimate of cost is invariably made by people who want to gain official approval for their plans, and who therefore include only the direct costs at the lowest figure which can be made to look plausible; the technical side of such costing is so complicated that the figures are difficult for anyone not directly involved to check, so that attempts by Government officials to do so are not always successful. When the project is well under way and the true costs have started to reveal themselves, cancellation may be more expensive than paying the extra. *Concorde* is perhaps the prime example, and deserves a book to itself solely on the history of the successive rises in development costs.

Estimates are sometimes made of the cost (or alleged cost) of something which will never find its way into any balance sheet. The costs of a delay, a strike, an accident or a natural disaster are frequently stated but, except in very simple cases, carry an air of unreality about them. In a strike, for instance, many cash amounts are affected: value of output, manufacturer's profit, strikers' wages, and the wages of other employees laid off are the most obvious ones. But many others, including customers, suppliers of materials, shareholders, tradespeople who serve the strikers and the Inland Revenue, will lose indirectly. To decide how many of these should be included, and how the figures should be calculated, requires so much subjective judgement that the result is unlikely to mean very much, even if there is no intention to mislead. One figure commonly quoted as a measure of the cost of a strike is the value of the output lost, but this is inaccurate in many instances because items held up during the strike are finished after it is over. Various other snags in this type of calculation are mentioned below.

Whether the figure is an estimate or a final and unalterable amount, a number of questions have to be asked.

First, is the cost a once-for-all sum or a recurrent figure? The types of commitment involved are very different. Also, if the cost is for an item of capital expenditure, will it produce an annual return of revenue or generate further annual expenditure? Detailed interpretation is a matter for economists, but the distinction is a fairly basic one. The Channel Tunnel again provides a good example, in which the decision to go ahead has to be based as much on the economics of its operation as on the capital cost—but usually it is only the construction costs which get the publicity.

Secondly, what does the stated cost include? The comparison between transport costs by road and by rail is notoriously unfair to the railways, because they are responsible for the whole cost of the service while the cost of road construction and maintenance is shared between the user and the taxpayer.

Thirdly, who benefits? If the choice lies between an imported item and a slightly more expensive home-produced one, it may be in the nation's (if not directly in the buyer's) interest to pay the extra, because it provides employment and saves foreign exchange. The indirect returns through taxation also come into the calculations.

Fourthly, who bears the cost? A multiple road crash costs the community a considerable amount in terms of physical damage, medical care, police time, traffic delays, loss of earning potential of those involved, and so on. But some of this amount is borne by the people in the crash, some by insurance companies and some by public funds, while to the vehicle manufacturers and repairers it is actually a source of income. To give an overall figure without taking such matters into account is misleading.

This is particularly true in assessing costs of social services. It is possible to make out what appears to be a good case for a free contraceptive service, for instance, by adding together all the savings which would result from a reduction in the number of babies said to be 'unwanted'. Quite apart from the guesswork involved in assessing numbers and costs, claims of this kind also tend to ignore the question—cost to whom? This particular one also involves an assumption that is just those babies which are said to be unwanted whose birth will be avoided by the measures proposed, which is by no means self-evident; it also ignores the fact that there is a large and mainly unsatisfied demand for babies by prospective adopters. To discuss such matters only in financial terms is much too simple.

Another example is the change in the United Kingdom from Imperial to SI (Metric) units, for which costs have been quoted both by the Government and by industries and firms concerned. But the estimates (at least those published in the press) do not reveal who bears the costs, and what proportion of them would have been incurred anyhow, perhaps later in time. So the figures tell the reader very little.

The timing of payments is also a factor in assessing true costs. If the cost of a certain project is £10 000, spread over four years with equal amounts to be paid at the end of each year, then the cash in hand required now is only about £8670, if interest at 6 % is allowed.

Allowance for inflation is an obvious enough point. But currency values can cause confusion in other ways, on account of variations in exchange rates. The rates themselves may be misleading in cases where there are legal restrictions on the free exchange of money. Comparisons of costs between one country and another therefore require care in interpretation.

The same holds true for other quantities which are expressed in currency units, such as Gross National Product and *per capita* income. The GNP is the sum of the incomes before tax of a country's inhabitants, and this divided by the population gives the *per capita* income. But these figures are independent of non-financial income so that, in subsistence farming areas for example, they do not give a fair estimate of a country's wealth. Again, comparisons between countries can only be made in very general terms because of the restrictions on the free exchange of money and goods.

The distinction between direct and indirect costs should always be taken into account. A fire at a department store causes losses of both kinds: direct losses through the damage itself, and indirect losses because business is taken elsewhere during rebuilding. For the individual, this is most important to remember when deciding the extent of insurance cover required in policies covering fire, accident, injury and theft.

The difference between an average cost and a prime or marginal cost is frequently overlooked. Suppose, for example, that over a full year a car costs 5p per kilometre to run—the average cost, that is, covering maintenance, tax, insurance and depreciation as well as fuel. This does *not* mean that if you walk 1 kilometre to a shop and back again, rather than use the car, you save 10p. In fact, you save the the cost of fuel, and arguably a little on maintenance, but that is all. This cost per *extra* kilometre is the marginal cost. The same principle holds in industrial production: in assessing whether output should be increased, it is the marginal cost and not the average which should be taken into account.

The concept of a notional value—an amount decided for purposes of planning and discussion, not because the actual currency amount could ever be exchanged for the item concerned—is not a new one. A similar line of reasoning has been used for many years in assessing legal compensation for injury and death; it is understood by all those involved that the word 'compensation' is a misnomer in the great majority of cases, but the concept has to be used in the absence of anything better. However, the idea has recently come into use in a more positive sense, under the title of cost-benefit analysis; the first major planning operation in which it was employed was to choose between competing proposals for military equipment in the United States. In this and in similar work elsewhere it has proved valuable.

But when the same technique has been tried out in attempts to compare the costs of things which are different in nature, the results have been less happy. To give a simple example, suppose a choice has to be made between two routes for a short length of motorway. One cuts through a housing estate, the other crosses a cemetery and would mean the loss of a thirteenth-century farmhouse listed as of historical importance. Which route causes more loss to the community as a whole, measured in financial terms? Notional costs for such matters as disturbance and noise, as well as historical and architectural merit, are difficult to judge fairly, and any costing method for a cemetery is bound to be controversial.

The principal danger in carrying out this kind of exercise is the risk of attempting to judge between two or more total figures differing by, say, 10% or 20%, when the various amounts making up the totals have been estimated on such vague and subjective criteria that the choice is meaningless. This was so in the case of the choice of site for the third London Airport, when figures of astronomical size seemed to the casual reader to show that the cost of the Maplin site (cost to whom?) would be about £100 million more than that of Cublington, the alternative inland site. Fortunately, in this case, the fallacies of the costing were seen in time.

A further danger in cost-benefit analysis is that matters which are even less quantifiable than the above may be overlooked in the mass of figures: danger to life, pollution outside the area directly involved, the effects on wild life and so on. An example quoted in an official report is that it would be quite possible to argue the merits of slavery on a cost-benefit basis: an extreme case, perhaps, but one that provides a fair analogy for others nearer home.

Perhaps in no other type of statistic is it possible to be at once so vague and so plausible as in matters of cost. As a result, both governments and individuals have made expensive mistakes. Only by persistent questioning can we hope to avoid them.

Drawing conclusions

Having checked the definitions of terms and the soundness of the figures being used, we must now find out whether they justify the conclusions being drawn from them.

The first point to look for is the use of statistical generalisations which are not based on figures at all. Common examples are statements that standards of morality (or personal service, or nutrition, or even cricket) have declined or are declining. The hottest and least productive debates at this level seem always to be on educational matters: of course, standards are changing but, since testing and assessment methods are changing also, as well as many other things in education generally, the figures which are available simply do not lend themselves to easy generalisations.

Appeals to public opinion fall into the same category. Here, the facts are usually easier to check, by reference to published opinion polls. But often such claims concern, not the balance of public opinion, but the direction in which that balance is changing—and this is more difficult to justify. Since the person who makes the suggestion usually holds the view which he claims to be growing in favour, he may be trying to guide public opinion rather than to estimate it.

Perhaps the most obvious of all the ways of misusing statistics is to use them selectively. When both sides of a debate are presented together, it is only to be expected that each side will quote the figures which best support its own case. It is when there is no one to put the other side of the argument that caution is necessary. So, in reading advertisements which rely on statistical argument, and in any other such one-sided situations, check to see what has been left out. Perhaps the commonest example is in advertisements for financial loans, in which the rate of repayment is stated but not the rate of interest, and sometimes not even the repayment time.

Particular vigilance on this point is required when looking at official planning proposals. The figures on which these are based are invariably selective, because they are primarily concerned with their public rather than their personal consequences. Thus it was possible for someone from the official side to suggest that London's proposed new motorway system was worth the destruction of 30 000 homes. Perhaps it was true, in terms of cold economic planning. But the statistics in human terms were rather left out.

A good statistical report often states possible objections. If the case presented is a strong one, the effect will be to strengthen it still further: it is the advocate of the weaker case who has to try to strengthen it by using the figures selectively. An excellent example is the MIT/Club of Rome's much-maligned but devastating report on the predicament of mankind, *The Limits to Growth*. Most of the criticisms directed at it were in fact anticipated in the report itself, and it is not something which can be judged by its reviews only.

Frequently, conclusions are stated in the form of a comparison, and the first point to check here is whether the figures are comparable at all. One of the best-known types of false comparison, that white sheep eat more than black ones, is still common: a newspaper article suggested that the number of deaths in mining accidents did not prove that the work was excessively dangerous, because eighty times as many people were killed on the roads. Whether mining is dangerous is a subjective matter on which a difference of opinion is quite in order, but this comparison tells us nothing useful whatever on the question: more people were killed on the roads because (to use insurance terminology) more people were at risk.

In a changing world, very few comparisons can be made without some objection being raised. Where objections are made, the total effect of the errors must be assessed in relation to the observed margin of difference. Thus if the *per capita* incomes of two countries are quoted as $200 and $250, problems of definition, measurement and currency exchange could account for the difference. If the figures were $200 and $800, this would be much less likely.

When a comparison is made between an observed figure and some reference standard, the standard itself should not be accepted uncritically: it may be too stringent, too liberal, or simply the result of guesswork when accurate information is not available. 'Safe limits' of contamination, both chemical and radioactive, have to be based on guesswork in some cases, because proper tests could only be carried out at great danger to life. In such matters, whatever tests may have been used to derive the figures, official bodies tend to err on the safe side: for instance, the nutrition standards of the Food and Agriculture Organisation are probably rather higher than actual minimum requirements.

Sometimes figures concerning one kind of measurement are used to draw conclusions about another. One example is the testing of a driver's blood alcohol level, either directly or by a breath test, when what is actually required is an estimate of the effect of the alcohol on his reactions to stimuli. Here the degree of association is high but by no means total, so that a light drinker whose alcohol tolerance is low may be unfit to drive when below the legal limit, while a more hardened one may be little affected by a level above the legal limit. Perishable foods, likewise, have to be assessed for their fitness for consumption, and the most convenient way is to set a time limit after the date of manufacture. But it is not an accurate measure, since other factors such as temperature also affect the condition of the food.

These are examples of correlation—that is, a quantitative relationship between two types of measurement. Another and better-known one is the connection between smoking and lung cancer. This is a big subject, full of problems which are difficult to explain without mathematics. But the main point is that correlation does not by itself imply and causal relationship: proof of cause-and-effect by statistical methods alone is very difficult. The subject is discussed in slightly more detail in Chapter 10 and, in diagram form, in Chapter 5.

The use of orderings can hide a multitude of fallacies. A description such as 'the biggest XYZ in the world', or 'the country with the third highest divorce rate in the world', should not be accepted at face value. A choice example is a sweet manufacturer's claim that he sells 'the longest-lasting fruit gum in the world'. One may reasonably assume that not all the various types have been tested, even if such a test could be objectively devised.

More seriously, an ordering does not always reveal the criteria by which one item is to be judged ahead of another. The world's biggest tree, for instance, might be claimed on a basis of height, girth of trunk, or estimated weight or volume. And the order does not reveal the magnitude of the differences between items, which may be so small as to be quite unimportant. In such cases, minor errors and variations in definition may be enough to upset the ordering considerably.

Next, some briefer points. Have exceptional figures been quoted as typical, or the other way round? Is there any appeal to scientific or medical authority? If so, check to see what has been left out: it may be quite true that 'doctors recommend' a particular medicine while being equally true that they also recommend its competitors, if they think there is little to choose between them. It is also worth asking how many doctors, and which ones.

Popular presentation methods, both graphical and pictorial, have also been mentioned elsewhere, in Chapter 5. Standards have improved here in recent years, but it is still possible to detect the occasional deceitful graph; one example was quoted in that chapter.

One final point: think whether the result is in accordance with reasonable common sense and, if not, be very suspicious. A survey of the work done by housewives suggested that, in view of the hours worked and the many different skills required, they were worthy of a high salary for it. Quite apart from all the possible errors involved in making such an estimate, what the survey omitted was that, if someone had to pay for all that work to be done at the market rate, it would never be carried out in the way that the typical housewife does it.

This is not to say that common sense is to be preferred to statistics, when the two seem to conflict: very often it is not. What one must be sure about is that the statistics are accurately obtained, and objectively used.

10 Introduction to some further statistical methods

As the heading indicates, the topics mentioned in this chapter receive only a brief mention. The intention is mainly to show what is available, so that readers who are interested in any particular method may follow it up in other books.

Significance tests

It sometimes happens that a pre-election poll of voting intentions shows a lead of one party over the other, while another organisation publishes a poll at the same time with quite the opposite result. Which are we to believe?

There may, in fact, be nothing wrong with either conclusion. Because each relies on a sample, the results are subject to sampling error, as mentioned in Chapter 6. So the difference between the two figures quoted for the proportion of voters supporting one particular party may be no more than can be accounted for by these errors. But there comes a point at which this is no longer so—when the difference is too large to explain away by the routine variations due to chance.

Except in extreme cases, common sense is not a good enough guide to the whereabouts of the boundary line. The technique required is called a significance test, and a difference of the kind given in this example is said to be 'significant' when it is greater than could reasonably be accounted for by chance variations. A conclusion that the difference was 'significant at the 1 % level' would mean that, if the two polls were each repeated a very large number of times, using samples drawn from the same populations on the same system as before, a difference as great as the observed one would occur less frequently than 1 % of all cases.

However, a statement that a result is significant is not the same as a statement that it is of practical importance. This is because the measure of significance depends on the size of the samples used, which is unrelated to the importance of the conclusion. If the poll organisers had each used samples of only a few hundred, an observed difference of, say, 6% would not be statistically significant, but it would be of great practical importance because it makes an accurate prediction of the election result impossible.

If a difference is shown to be significant, the next step is to seek the cause. For the poll figures, one or both of the sampling procedures might have been at fault, or the samples might have been taken from different populations, either from different areas or using different frames.

As an example of another kind, consider an experiment designed to test a new drug against an established one. A sample of 200 patients was divided randomly into two groups of 100; of those given the new drug, 63 showed an improvement within three days, while of those given the other only 49 did so. Here the cause—if it is the true cause—is obvious at once, and the aim must be to exclude the possibility that the difference might reasonably be due to chance. The significance level in this case is about 5%, meaning that if one gave the *same* treatment to two randomly chosen groups of 100 patients, the probability of a difference at least as great as the one actually observed is about 0·05. A medical researcher, therefore, might be unwilling to stake his reputation on a claim that the drug was a real improvement: there is just this chance that he might be wrong. He would be well advised to carry out further tests before publishing anything.

Conventionally, a figure between 5% and 1% is said to be *significant*, between 1% and 0·1% *very significant*, and less than 0·1% *highly significant*. The lower the percentage, the more significant the result.

The significance test is closely linked to the idea of a confidence level, which was mentioned briefly in Chapter 6. Suppose that a coin is thrown 200 times and 130 heads appear: this suggests the possibility that the coin is biased. To test this, we could carry out a significance test, which would give a measure of the probability of obtaining, with a fair coin, a result as far from the score of 100 heads as the observed number of 130 heads—100 being the score which would be expected on a basis of chance. This probability is in fact very small, and so there is a strong indication of bias.

Alternatively, we could calculate confidence limits for the true value of the probability of a head, given the experimental result of 130 heads in 200 throws. The calculation shows that the lower 99·9% confidence limit is above 0·5, meaning that we can conclude, with a high degree of confidence, that the coin is biased. Each of these two tests gives more or less the same conclusions, expressed in different words.

The chi-square test

Suppose that an ordinary die is thrown 600 times, and that the numbers of each possible score, in order, are 93, 86, 121, 91, 127 and 82. The figures indicate that the die might be biased, as scores 3 and 5 appear far more often than the others. But, once again, are the differences too great to attribute to chance?

The basic idea is similar to that of a significance test. In fact, we could apply a significance test directly to the proportion of any one of the scores. But, in order to test all the scores together, as is necessary here, a further test is required. This is the chi-square test, so called by reason of the symbols used in the necessary mathematical formulae, χ^2; this Greek letter is anglicised as chi, and pronounced ky. The conclusion drawn from the test is expressed in the form of a significance level; in this case, the level is under 1 %, giving a strong indication that the die is biased.

The chi-square test can be used whenever the results of a statistical experiment have to be compared with predictions made on a basis of probability, and whenever two or more sets of results of different experiments have to be compared with each other. Thus the scope of the test is very wide, and it is particularly useful in medical work.

Correlation and regression

The concept of correlation, or association, between two sets of figures has been mentioned briefly in earlier chapters, and no further explanation of the basic idea is given here; a few supplementary points, however, may be useful.

The measure used to assess the degree of association is the correlation coefficient, which is measured on a scale from 0 to 1. Zero represents no correlation, while 1 represents a direct relationship in which the value on one scale can be used to predict the exact value on the other. If plotted on a graph, with separate scales for the two sets of measurements being compared, the points lie on a straight line if the coefficient equals 1, and fairly close to a straight line if the coefficient is close to 1. (See Figures 13 and 14 in Chapter 5.) If increasing values on one scale are associated with decreasing values on the other, the coefficient is negative, so that a value of —1 represents a direct relationship of this type.

The interpretation of the actual values of a correlation coefficient is difficult, and best left to those familiar with the technique involved. The main reason for the difficulty is that the significance of a particular value depends to a large extent on the number of measurements.

In certain types of test, objective measurements are impossible and the results are given in terms of an order of merit or preference. Examples are the kinds of competition often found on breakfast cereal packets. When two orders are given by different people, some association may be expected between them, and a further type of correlation coefficient can be used to measure it. Since the positions in order are known as ranks, it is called a coefficient of rank correlation. Two versions of rank coefficient are in use, named after C. Spearman and M. G. Kendall; as before, interpretation of actual values is not for amateurs.

Rank coefficients may be used in correlating any two sets of figures if they are put into rank order first: for example, comparing orders of merit for, say, French and German in a school class, rather than comparing the marks on which those orders of merit were based. But, since an order of merit fails to distinguish between wide margins of difference and narrow ones, the results obtained will not be quite the same.

When the correlation is very close, it is possible to predict values of one variable from values of the other. Suppose, for example, that an examination was in two parts, and that a candidate was ill for one part. It would be possible to assess, on a basis of the comparative marks gained by the other candidates on the two papers, roughly what mark should be credited for the missing paper. Of course, there is a measure of uncertainty, but the result obtained would be better than just giving an average, or indeed than any other possible method of assessment. The technique involved is called regression analysis.

A practical example from engineering is worth mentioning. For certain types of steel, there is quite high correlation between the ultimate tensile strength and the hardness. It is usually more important to know the strength than it is to know the hardness, because the result of low strength may be a fracture; but hardness is easier to measure, because the test is non-destructive and requires less equipment. If it is known how the correlation operates, it is possible to predict, from a given hardness measurement, the corresponding

154 Understanding Statistics

ultimate tensile strength—as usual, within certain limits rather than as a precise figure. An approximate value, found quickly and cheaply, may in some circumstances be more appropriate than an accurate, slow and expensive one.

Statistical quality control

In a manufacturing process, there is usually a limit specified for each principal dimension in the form of a tolerance. If a dimension is outside the stated limits, the article has to be rejected. But this as it stands is rather a crude way of ensuring a satisfactory output, being both slow and wasteful. A great improvement may be made by keeping track of the routine small variations in the measurements, even when these are well within the tolerance. A typical arrangement is as follows.

Suppose a dimension has to be in the range 60 ± 1 mm. A sample of five items is taken, and the measurements are recorded for each. If one of these is outside the tolerance, action must be taken at once, but the whole point of the operation is to avoid such a situation. The mean of the measurements is then found: this has the effect of smoothing out the routine small variations from item to item so that, if everything is going normally, the mean should be very close to 60 mm.

But if the process is going wrong—for example, on account of wear in a cutting tool—this will have a visible effect, because it will tend to bias the errors in one direction. A mean of, say, 60·15 mm (or 59·85 mm) might therefore indicate that the situation needed watching, even though every item produced was still within tolerance. Such a value is called a *warning limit*.

Further samples may show that a correction to the machine setting is necessary, or alternatively may indicate that it was a false alarm. The results are usually plotted graphically, so that any trend becomes visible at once. Outside the warning limits (usually three times as far from the mean, or 60 ± 0.45 mm here) are the *action limits*, whose name is sufficiently descriptive. This form of quality control is widely used in industry.

Linear programming

Consider a lathe which can be used for producing two types of component, spindles and bolts. Each spindle takes 3 minutes, and each bolt takes 5 minutes. How many of each can be produced in an 8-hour shift?

If the lathe produces only spindles, 160 can be made; alternatively, the maximum number of bolts is 96. If some of each are required, the limits have to be expressed algebraically: the number of spindles is denoted by x and the number of bolts by y. Since one spindle takes 3 minutes, x spindles take $3x$ minutes; the bolts correspondingly take $5y$ minutes, and the sum of these must be less than or equal to 480 minutes. More briefly, $3x + 5y \leqslant 480$.

Any combination may now be tested. If 45 spindles are required, $3x = 135$, so that $5y$ cannot be more than 345 if the total is not to be more than 480. Hence the largest possible number of bolts, made together with 45 spindles, is 69.

This is known as a linear inequality, or constraint, in two variables—linear, because a graph of the values of x against the corresponding maximum values of y is a straight line. In a full-scale industrial process, there may be many such constraints: availability of equipment of various kinds, manpower, space, sales potential and capital. Also, there may be more than two variables—the quantities of each different product being made, for example. The technique required to find the optimum value of each variable, in such a way that the whole process works at the highest possible efficiency, is called linear programming. It is well suited to analysis by computer, and in simple forms is now being taught in schools.

Critical path analysis

To prepare a cup of tea for two, first get out the cups, sugar and milk, put the tea in the teapot, then put the kettle on and wait for it to boil. Is this the best possible way? Of course not, because if the kettle were to be put on first all the rest could be done while it was heating up. Since the heating of the water takes the longest time, and can be started without having to wait for any other part of the operation, it should receive priority. Further, the time taken for it to boil determines the time necessary for the whole process. In planning terms, with the separate parts of the operation shown by arrows on a chart, the boiling of the kettle is the 'critical path' from the start of the process to the making of the tea. If something goes wrong—no sugar available, so a dash to the shop to buy some—the critical path may change.

This is not a difficult example, of course. As a next step, the reader may like to work out the critical path for the preparation of a three-course meal, in which some operations (cook potatoes) cannot be started until others (peel potatoes) have been completed; also, some pairs of operations cannot be done simultaneously because they both require the use of the available manpower, while others can be left to get on by themselves once they have been started.

At this level of difficulty, critical path analysis is mainly an instructive intellectual exercise and, as such, it is finding its way into schools. But it really becomes useful in situations of much greater complexity. Consider the problem of the construction of a steelworks, in which many different types of job all have to fit together in sequence. The efficient planning of such an undertaking is difficult enough in itself, but to decide what action to take when parts of the process get behind schedule is harder still, and this is where CPA shows its real power. Once again, a computer can be used to do the analysis.

First, a list is drawn up of every operation, the time it takes and the operations which must be completed before it can start. The computer will then work out the critical path, and the earliest and latest starting times for each operation. For something in the critical path, these two times are the same, and a delay holds up the entire schedule; for others, there is some freedom of movement, and considerations such as the efficient use of manpower and equipment will help in choosing the optimum for each. While the work is in progress, information can be given to the computer each day to provide an up-to-date picture of the situation, taking account of work completed, changes in manpower availability, late deliveries of equipment and other delays, and even changes in the weather.

Neither CPA nor linear programming is usually thought of primarily as a statistical technique, but this is what they both are; no list of the statistical methods which are of most practical use would be complete without them.

11 Top of the charts

Statistical methods are of fundamental importance in the science of market research, which is widely used by commercial firms as the best way of matching their products to the requirements of those who buy them. However, since the information which is most useful to a firm is the information which its competitors do not have, the results of market research work are rarely mentioned in newspapers. One of the few cases which do reach the notice of the public is the popularity chart of current records which is used in the BBC television show, *Top of the Pops*. Very few of those who watch the programme, perhaps, give much thought to the statistical work which lies behind the chart. But the process is by no means straightforward, and quite instructive for several reasons.

There are many possible ways of assessing popularity in this context. One is to estimate the total sales of each record; another is to estimate the number of times each one is played; another is by questionnaire, using either a changing sample or a panel; another is to use indirect evidence such as letters to radio request programmes. The requirements of the method to be chosen are fairly obvious: it must be quick, accurate, not too expensive and, as far as possible, proof against manipulation. Also, it is necessary to decide what exactly is meant by the word 'popularity', and then choose a method which measures just this and not something else.

The idea of using the number of radio requests can be discounted at once, because manipulation is easy to carry out and difficult to detect. This cannot be allowed to happen, since the chart not only reflects the state of the market—it also helps to guide the movements within the market. Since large sums of money depend on these movements, any opportunity for manipulation of the charts will certainly be taken up.

The use of a panel allows too much scope for subjective choice by the researchers, in selecting those who are to form the panel. This would give rise to disagreements from people whose musical tastes were different from those of the panel. Also, the panel members would soon become 'conditioned', and the opportunities for outside influence would be great.

A sample survey of consumers would give satisfactory results if it could be carried out. But the task would be too big and too slow, and probably too expensive as well because only a very small proportion of the whole population buy a record in any given week. There would be an obvious temptation to simplify the operation by restricting the sample to those of the age range primarily concerned, but the effect would be to introduce bias into the result.

The information needed to work out how many times each record is played is simply not available, and could not possibly be found even by using a sample survey. This is perhaps the method which gives the best measure of true popularity. But, since it is impracticable, one must look for an alternative, and the only method which remains is the estimation of sales. This can be carried out relatively cheaply; it is quick and fairly accurate; and although manipulation is possible, attempts can be detected.

But it should not be assumed uncritically that the number of copies of a particular record which are sold in a short period of time is a true measure of its popularity. In assessing the popularity of children's books, for example, such a measure would not be accurate, since many of these are bought by adults whose tastes are often quite different from those of the children who receive them. However, the sales of records give a much better guide, as only a small proportion are bought as presents.

Also, the world of pop records is a world of ephemera, so that few records bought a decade or more ago are still in use now. If there were many of them, their current popularity would not be reflected in the sales figures.

Perhaps the biggest error is caused by the fact that the pattern of purchases differs from one record to another. Thus a record which maintains a steady but unspectacular sale over a long period of time does not reach the charts, while one which has a high sale for just a few weeks can go high in the charts and yet sell fewer during a full year. But which of the two is the more 'popular', even during the peak selling period of the ephemeral one, is in any case a debatable point. The net effect of such variations is probably quite small, bearing in mind that the chief interest in the charts is centred on the short-term changes which they reveal. So, whatever its minor imperfections, the weekly sales chart is the best available indicator of popularity.

The research work for the BBC chart is carried out by the British Market Research Bureau Limited, and the results are also published in the magazine *Music Week*. A panel of 300 record shops is used, with the correct proportion of shops of different sizes and types, and with every region of the country properly represented. The panel covers about 7% of the total number of record shops in Great Britain.

Each shop on the panel makes a list on which the serial number of every record sold (both singles and LP's) is entered at the time of sale. The list for each week is sent off to BMRB after the shop closes on Saturday. The lists are checked for suspicious signs such as one serial number written several times in succession, and the blank spaces are crossed out to avoid unauthorised additions.

When the check is complete, the lists are sent off for processing by computer. The print-out gives two charts for the week, for singles and for LP's, in descending order of sales. At this point the precautions against manipulation (or hyping, as it is called in the record business) are taken, as part of the routine.

Although the identities of the shops on the panel are not published, it is not practicable to keep them entirely secret. It is possible, therefore, for someone to buy (and encourage his friends to buy) copies of a particular record at a shop or group of shops known to be on the panel. A statistical test to guard against this possibility is carried out whenever a record enters the singles chart for the first time, and whenever a record rises ten or more places in the singles chart between one week and the next. The test involves taking a random sample of fifty shops from those not on the panel, and telephoning each one to ask about the sales of the records concerned.

If doubt still remains, the record may be excluded from the chart as published, and a full enquiry follows. Since any shop which has been involved (knowingly or unknowingly) in an attempt at manipulation can be identified at once, such attempts carry a high probability of detection. Also, if the attempt is made without the connivance of the shop, it is an expensive operation to carry out because the records have to be bought at the retail price.

The total sales figures for each record, over a long period of time, can be estimated by scaling up the figures obtained from the shops on the panel, and a comparison with the manufacturer's own sales figures is then possible. No exact agreement is to be expected, of course, for three reasons: sampling errors, the time-lag between the two stages of the sale and the fact that stocks held by the retailers are left out of the calculations. But such information as is available along these lines has confirmed the substantial accuracy of this type of chart.

12 Statistics in medicine

The use of statistical data and analysis is an integral part of medical work—diagnostic, curative, preventive and even administrative. Four short illustrations only are given here, as a brief introduction to a wide-ranging and fascinating subject.

First, an example from the past. The cholera epidemics of the last century, particularly that of 1831–3, led to a great deal of research in order to find the cause of the disease, and the cholera bacillus was finally discovered in 1884. But, long before that, the means by which the disease was transmitted had been found, and this probably made a greater contribution to the successful control of cholera than the later discovery of the true cause.

One of the researchers, Dr John Snow, drew maps of the area he was studying, Southwark in London, to show where each person contracting the disease lived. He was not the only one to do this, of course—another researcher who did the same noted that the height above sea-level seemed to be connected with the incidence of the disease, and he formed a theory on this basis. Snow had a similar idea, that the incidence was related to the supply of drinking water. He concluded that something carried by the water was responsible. As a piece of statistical analysis, it was not particularly advanced, as it depended only on the keeping of records and on a common-sense interpretation of the data. It was the care and thoroughness with which this work was done which eventually proved conclusive.

His theory was first published in 1849, and did not find immediate acceptance. But he continued with the research and eventually, in 1854, the opportunity for a conclusive proof presented itself. He was investigating an outbreak centred on Broad Street in London—now Broadwick Street, Soho. The problem was to show that the statistical association between the cases of cholera and the water supply was the only kind of association which existed. And the thoroughness of his investigation revealed several cases in which this could be demonstrated.

The most unmistakable one was that of a widow, who at first sight seemed to have nothing to do with the Broad Street outbreak at all, since she lived in Hampstead. But Snow discovered that she had formerly lived in Broad Street, and that while there she had developed a liking for the water from the local pump—so much so that she had arranged for a regular supply to be sent out to Hampstead for her. On August 31st some was sent; on the following day she caught the disease, and on the day after that she died.

The evidence from this and other cases was now unchallengeable. Snow's action was dramatic—he persuaded the local Board of Guardians to remove the handle from the Broad Street pump. Within days the number of cases had dropped to almost none. It was a turning-point, not only in the conquest of cholera in England but also in the history of public health.

Next, a medical-statistical controversy which has lasted for thirty years and, as yet, shows no sign of abating. In the early years of this century, a form of tooth decay observed in Italy, England and the United States was investigated, and by 1931 it seemed clear that contamination of drinking water by various fluorides was responsible. But the research showed one unexpected result: that, although large amounts of fluoride were harmful, small amounts actually strengthened the teeth, especially during childhood.

Further research, particularly in areas where fluoride occurred naturally and found its way into the water supply, indicated that about 1·0 parts of fluoride ion per million parts of water was the optimum—it strengthened the teeth as much as possible while apparently having no ill effects. (The actual figure varies slightly, being rather less in hotter areas where more water is drunk.) In 1945, the time had arrived for a controlled test to be carried out.

Since drinking water is distributed over a wide area from a single source, an experiment using two random samples of individuals was impracticable. The next best thing was to take separate samples from two towns which were as closely matched as possible, and to add fluoride to the supply of one of them. This was done, in fact, in three pairs of towns, and in two cases a third town with natural fluoride was used as a further control. It will be sufficient to describe one of these cases here.

Grand Rapids, Michigan, with natural fluoride at the low level of 0·15 parts per million (ppm), was selected for fluoridation up to 1·0 ppm, and Muskegon in Michigan, also with 0·15 ppm, was the main control. The town of Aurora in Illinois, with a natural fluoride level of 1·2 ppm, was used as the second control. Samples of children of various ages were selected from each, and given thorough medical and dental examinations before the fluoride was added. When the three sets of results were compared, no significant differences could be found except that the number of decayed, missing and filled teeth (known as the DMF rate) was less at Aurora than at the others. Fluoridation was then started, and the examinations repeated at intervals for over ten years.

Within a few years, the results were impressive. The DMF rates in Grand Rapids showed sharp improvements, and for the children born during the fluoridation period (who were added to the samples as the experiment went on) the rate approached that of Aurora, a reduction of about 50% on the earlier figures. Older children showed progressively less improvement. No adverse results were detected, though many people hostile to the experiment were quick to allege side-effects of various kinds. Every such suggestion was tested, but none was shown to have occurred. Most authorities in this field would agree that this is still substantially true today, though a few doubtful points remain: for example, the possibility that people with medical conditions requiring them to consume large quantities of water may receive excessive amounts of fluoride.

No one, even now, knows *why* fluoride strengthens teeth and reduces decay. The evidence is statistical, and the medical aspect has been confined to the control of the fluoridation process and the monitoring of its effects. But the man in the street shows a strange reluctance to be convinced by statistical argument. It may be because statistical methods are so widely misused, but the real reason seems to lie deeper in the human personality—it is, at any rate, far outside the scope of this book. Whatever its nature, it had produced, in every country and district where the addition of fluoride has been proposed, anti-fluoridation groups who have been zealous in discovering every possible legal, medical and moral excuse for refusing (on behalf of their children, for adults are unaffected) the privilege of healthier teeth. The final outcome of the controversy seems a long way off, and clearly depends on more conclusive medical evidence than is available now.

The third example is in some ways rather similar: the association between smoking and lung cancer. The fact of the association is both firmly established and well enough known, and it is only a single aspect of it which will be mentioned here.

Although it is beyond dispute that those who smoke heavily have a much higher death rate from lung cancer than those who do not smoke, it cannot be deduced from this that smoking causes lung cancer. The samples being compared are not randomly selected from the whole population, and two other explanations are possible: either incipient lung cancer causes people to start smoking, or the kind of people who smoke also have certain qualities which tend to increase susceptibility to lung cancer. After the publication of the first Royal College of Physicians report on Smoking and Health in 1962, these objections were made.

But the difficulty had been foreseen by two of the doctors responsible for the earlier research, Dr Doll and Dr Hill. In a paper published in 1964 they produced a masterly answer to the objections, which removed the only remaining way for a heavy smoker to persuade himself that the noose round his neck was not really there.

The method they used was to trace the history of a number of people who had stopped smoking. If their lung cancer rates had continued to be typical of smokers, no conclusions either way would have been possible. But, in fact, the rates dropped sharply, to less than half within five years and to about a fifth after fifteen years. (The rates were standardised to allow for the different ages.)

Thus both incipient lung cancer as a reason for people starting to smoke and an independent third cause were ruled out. A fuller account will be found in the second RCP report, *Smoking and Health Now*, published in 1970, which covers many detrimental effects of smoking as well as lung cancer. A copy of it should be kept in a prominent position in the house of every cigarette smoker, as a devastating exposure of the statistical (and medical) evidence he is choosing to ignore.

Finally, an example of a simple piece of statistical analysis which shows that common-sense interpretation of numerical data can sometimes be misleading. The problem is that of screening—carrying out a simple test on large numbers of people in order to detect cases of a particular disease, such as a form of cancer, at the earliest possible stage.

Since the test has to be a simple one, it cannot be absolutely foolproof. It will give occasional positive indications where the disease is not present, and will allow some actual cases to escape detection. Suppose the success rate is 99 % in each case—this is probably a rather higher figure than would obtain in most such tests. One of the rates could be increased without difficulty, but only at the expense of the other: if all the doubtful cases were counted positive, more of the true cases would be discovered, but more negatives would be lumped together with them.

Suppose also that the incidence of the disease is 1 in 1000, and consider what happens when 100 000 people are tested. If the numbers correspond exactly to the proportions assumed, there will be 100 people with the disease, of whom 99 will show a positive reaction to the test. But, of the 99 900 who do not have the disease, 1 % or 999 people will in error show a positive reaction. So, of the 1098 who show a positive reaction to the test, only 99, or about 9 %, actually have the disease. Since the test is 99 % efficient, this seems an improbable figure, but it is quite correct.

The 1098 individuals would therefore have to be recalled for a further test. This time 98 out of the 99 with the disease would show a positive reaction, and only 10 of the 999 healthy individuals would show a positive reaction. Alternatively, the smaller numbers to be tested this time might make a more sensitive test possible, so that even these few could be eliminated and the one actual case not allowed to escape.

This explains why a postcard from the testing centre, asking an individual who has had a screening test to return for further tests, should not be taken as near-proof that the disease has been contracted. The usual practice is to state this on the card itself.

13 The measurement of personal achievement

Statistics, as by this time the reader will be well aware, is concerned as much with the precision with which a quantity can be measured as with the measurement itself. The distinction is nowhere clearer than in the matter of examinations. To award a grade 3 pass in 'O' level French is one thing; it is quite another to assert that the candidate would still have gained a grade 3 (or, at worst, a 2 or a 4) if he had sat for the examination of a different Board, or if he had a different teacher, or if he had been assessed on two years' course work instead of four hours' writing from memory. Since most readers either have been or will be in a position where such measurements are of vital importance, a brief look at the statistical problems involved seems appropriate.

One of the possible reasons for an error in any measurement is the lack of a single, objective way of defining the necessary scale. When there are two or more ways of choosing a scale—all of them, perhaps, useful for different purposes or simply preferred by different groups of people—confusion arises.

Suppose, for instance, that we are measuring the flow of water through a pipe, and that we have two meters for the purpose. Meter A reads 6·5, meter B (calibrated in different units) reads 72. Then we measure the flow through another pipe: meter A reads 8·2, meter B reads 59. Which pipe has the bigger rate of flow?

It seems that an error must have been made, since meter A gives the second pipe the larger figure, and meter B the first pipe. But what has happened is that meter A has measured the *speed* of the water, and meter B the *volume* of water per second. The first pipe was larger than the other so that, even with a lower speed, the volume passing each second was higher. Thus there were two methods, equally accurate, equally useful in their different ways, giving results which appeared to contradict each other.

By comparison with the multiplicity of methods available for conducting an examination, this example is a simple one. Each choice made, from the syllabus to the marking scheme, affects the interpretation of the results in the same way as does the choice of meter to measure the flow, and the cumulative effect of all these choices is very great.

From the point of view of someone who is trying to construct an examination which does what he wants it to do, the optimum choice is the one which best fulfils four principal requirements.

(i) **Reliability**. This means that the general pattern of results must be consistent between different times and different places, when used for the measurement of the same or comparable groups. A method which is shown to be reliable may have a different scale of measurement from another which is equally reliable, so that reliability does not imply uniqueness. (Both meters in the example above were accurate, so both were reliable.) Comparisons made between two sets of measurements are essentially statistical, since no exact correspondence is to be expected; the tests which can be used for this purpose range from the very simple and subjective (do the same people usually come near the top of the list in each of a number of similar tests?) to highly refined ones using computer analysis. An example of the latter is mentioned below in the section on multiple choice tests. A further example, suitable for either manual or computer analysis, is the alternate-answer check, in which the pattern of answers to the odd-numbered questions is compared with that of the even-numbered ones. Some precautions against bias are, however, necessary for this.

(ii) **Validity**. In other words, does it measure what we want to measure? For the flow measurement through the pipes, one method is valid and the other not, and one cannot say which without knowing the purpose of the operation. Suppose we have a series of tests which, when given in turn to a group of people, produce reasonably consistent (that is, reliable) results. We can then say that those who score high marks in one of the tests have what we might call a higher X rating than those with lower marks, where X is the particular ability which the test is measuring. What is *not* proved, however, is just what this ability is. It is not 'intelligence' merely because the test carries the label 'Intelligence Test'. Neither is it history, mathematics, theology or accountancy merely because the questions have been on these subjects. One hopes, of course, that it will be. But, to say the least, it is not universally so.

(iii) **Discrimination**. Do the results show a reasonably wide margin between the marks of those who do well and those who do badly, so that the chance variations inherent in any examination do not have too great an effect? (This is assuming, of course, that there is a reasonably wide range of ability in the candidates.)

(iv) **Standardisation**. If the results are to be interpreted by outsiders, or to be compared with the results of other tests, this is important. All the examination boards, for instance, must try to ensure that their standards are the same for a particular subject at a given level. Sometimes standards in different subjects have to be compared—as, for example, when fixing entry qualifications for universities—and this is more difficult. Standardisation is too big a subject to go into here—

it is full of problems, particularly in comparing standards over a period of time.

In all these requirements, an understanding of statistics is necessary if the setting and marking of an examination, and the interpretation of its results, is to be properly done. But experience is needed as well—it is not a subject which can be learnt from books.

The most obvious of many factors which affect the method of examining is the subject itself, together with the choice of syllabus within the subject. To give an example from each end of the scale, examining arithmetical ability is easy, and several methods could be used, while skill in teaching economics or carrying out surgical operations is extremely difficult to assess. In language teaching, grammatical correctness is easier to test than verbal fluency. In history or geography, an examination may be designed so that the candidate who has a good memory but little understanding does well or that he does badly, and the syllabus can be written with one or the other aim in mind.

Ease of testing often helps, in fact, to determine the syllabus. Language examinations, both English and foreign, have perhaps been influenced by this more than most, but others have been too—even mathematics, which is generally one of the easiest subjects to examine. One of the influences here is the difficulty of setting questions which are sufficiently unpredictable, with the result that test of understanding (as opposed to memory) cannot be constructed. As another example, the procedure for the standard driving test depends more on what *can* be tested than on what *ought to* be tested.

The relationship between the teacher, the examiner and the candidate is also important in personal terms. In a statistical enquiry by verbal questioning, the appearance, personality and manner of an interviewer affects the answers which are given, and the interviewer's judgement of the person similarly affects his interpretation of the answers in cases of doubt. In any examination other than a wholly impersonal one, the same thing happens—and to a much greater extent, because the result of the examination is normally of such major importance to the people involved. Even an examiner who tries to be fair may unconsciously penalise a candidate of scruffy appearance—or, conscious of a distaste which he does not want to translate into unfairness, he may over-correct to the point of generosity. A bias against women (or in favour of attractive ones) is also not unknown.

When the subject-matter is in any way controversial, such problems multiply: economics, politics, theology, musical interpretation and (once again) the driving test are some of the more obvious examples. In such cases, the candidate may be guided in his answer by what he thinks will be acceptable to the examiner, rather by his own opinions. An examiner may try to be fair and may even mark generously a well-presented exposition with which he disagrees, but it is a lot to ask of human nature.

The examination itself may be oral, written, practical or any combination of the three. Problems of standardisation occur in all but the very simplest of tests, though clearly they are greatest for oral and practical ones. In this brief introduction, only written tests will be considered.

The possible forms of written answer are the essay, the short answer (from a single word to a near-essay), and the multiple choice type. For mathematical and technical subjects some change of definition is required, but the same choice remains in principle.

Of the three, the essay form is the longest established and still has many defenders. But it lacks reliability for a variety of reasons, the main one being that markers tend to differ in their assessments. It also lacks validity, in all subjects other than English, because fluency in written English is being tested as well as the subject itself. Short-answer questions share these disadvantages to a lesser extent, but the main strength of the essay—that it, alone, requires the candidate to select the relevant material from his store of knowledge and to set it down in a clear and logical order—is in danger of being lost. It is this type of question which can often be replaced with advantage by the multiple choice type, to which we now turn.

Multiple choice tests

In this type of test, the questions themselves are similar to those requiring a short answer in an ordinary examination. But each question is followed by several possible answers (usually four or five), of which only one is right. The candidate simply has to choose the correct one. For example:

1. The parable of the Good Samaritan appears in the Gospel according to (*a*) Matthew; (*b*) Mark; (*c*) Luke; (*d*) John.
2. The square root of 0·0016 is (*a*) 0·004; (*b*) 0·04; (*c*) 0·0008; (*d*) 0·0126.
3. The Peasants' Revolt took place in (*a*) 1381; (*b*) 1389; (*c*) 1391; (*d*) 1831.

The advantages of this procedure are threefold. The marking becomes much easier, for a start, so that automatic marking can be used. This, in turn, means that there can be no errors in marking. And the elimination of discretionary marking encourages greater reliability.

It might be thought that handwritten single-word or numerical answers should not involve any doubts about correctness. But they often do: variations of spelling, and poor writing, may be marked either generously or severely.

When the number of candidates is too small to justify marking by computer, marking by hand is still much easier than for the traditional type of test. In an actual examination, a standard answer sheet is used, with spaces for each possible answer. For easy marking, the correct spaces are marked on one of these sheets, and then cut out with a sharp blade. When this sheet is placed over a candidate's answer sheet, his correct answers appear in the holes and can be counted very quickly. It is necessary to carry out a separate check to ensure that the candidate has not indicated two or more answers (unless this possibility is allowed for in the test), as this, of course, must not be permitted.

One commonly used system of automatic marking works like this. For each question, the printed answer sheet has the four possible answers shown thus:

[*a*]   [*b*]   [*c*]   [*d*]

To indicate the correct answer (*c*) for question 1, the candidate fills in the space between the square brackets with a soft pencil. To make a correction, a mark may be erased. In the marking device, the answer sheet is placed under a block of electrical contacts, one pair for each answer space. The graphite from the pencil completes an electrical circuit wherever a mark has been made, and the correct ones are counted up. When two spaces have been marked, no score is recorded.

The use of automatic marking makes it very easy to analyse the pattern of results for each question, which is an added advantage. Sometimes a question proves too difficult, so that most candidates guess the anwer: this shows up in the analysis, because the better candidates (measured on a basis of total scores) are no better at guessing than the others. It is a common practice for the results of questions which fail to discriminate to be ignored, and the marks revised accordingly. Since this reduces chance variations, it increases reliability. Such a process would be impracticable in a manual marking system. A better method of detecting such questions is to try them all out beforehand, but this is not always possible because of the need to preserve secrecy.

These advantages are not to be lightly ignored, and indeed have led some enthusiasts to bestow on this form of examination the tendentious label of 'Objective Test'. But only the marking is objective, and even that only on the assumption that all questions are fairly represented in the marking scheme, which is usually a matter full of subjective choices. The testing itself can no more be described as 'objective' than can any other kind of examination—in short, not at all. The alternative name of 'objectively marked test' is therefore better than 'objective test'—but 'multiple choice test' is both descriptive and accurate. The advantages remain but, before anyone hails the advent of a testing process superior to any other, some disadvantages must be pointed out also.

The most obvious one is the possibility of guesswork. In one way, this is not a serious difficulty, because a simple adaptation of the marking scheme can help to take account of it. The procedure, for a four-choice test, is to score 3 marks for each correct answer but subtract 1 mark for each wrong one. The candidate who guesses 40 answers, getting the expected number 10 of them right, scores 30 for these but loses 30 on account of the wrong ones, so that he does no better than the candidate who leaves the 40 answer spaces blank.

The important error resulting from guesswork is not this, but the false discrimination introduced by the chance variations in the guesses. In 40 guesses, not all candidates will get 10 correct—some will get 5 or less, some 15 or more, in numbers which can be estimated quite easily on a probability basis. And no amount of statistical manipulation can put matters right, since the deletion of a question as described above is only possible when a very high proportion of candidates guess in that particular question. Differences such as these can account for a considerable part of the whole range of marks and, in a pass/fail situation such as the notorious

11-plus, the consequences in personal terms cannot be dismissed as trivial.

A further disadvantage is that multiple choice tests lack full validity because a candidate using intelligent guesswork can often gain a high score, given only a limited knowledge of the subject. Consider, for example, someone who has never heard of the Peasants' Revolt (question 3 above). He sees that three of the answers are close together, and reasons that one of these is likely to be the correct one—the other two being wrong, but close enough to the true date to sound plausible. The fourth is also wrong, but this time designed to catch out someone who gets the order of the digits mixed up. Hence the correct answer must be 1381.

Of course, this is a badly set question, included here only to highlight the problem. But anyone who makes the attempt will soon find that it is very difficult to construct four answers which, considered as a group as well as individually, are equally plausible. One might almost say, looking at the need for the score obtained from guesswork to be no better than chance, that it is an impossible task: one has to choose a random sample of four from the population of all possible answers, with the restriction that one of the four is a non-random member of that population—the right answer. As a further illustration of the difficulties this may cause, consider question 2 above.

In this case, none of the four answers is an arbitrary choice. A combination of the correct answer (0·04) with three arbitrary figures would at once make the answer obvious, because of the connection between the 4 in the right answer and the 16 in the question. So at least one of the false trails must be plausible to the same degree as the right one—a further step away from randomness. In fact, all three are designed to allow for errors which might be made by candidates who were not shown the list of answers. One of these errors (getting the decimal point wrong) yields a wrong answer very similar to the true one, thus suggesting to the candidate that one of the two similar answers is correct and raising the chance of success by guesswork to 50%.

Secondary knowledge may also be used. A candidate who knew something of the background to the four gospels, but had never bothered to read them, might reason out question 1 as follows: the first three gospels have quite a lot of material in common, so that a passage which is only in one of the four is most likely to be in John. This time he would be wrong—but the aim is to improve on random guesswork, not to score 100%.

A more obvious feature of multiple choice tests is that the printed answers serve both to jog the memory and to exclude answers not stated. Whether these are disadvantages depends at least in part on the extent to which the examination is intended to be a test of memory; if it is primarily a test of understanding, a well-designed question and sequence of answers should cause few difficulties on this score.

Equally obviously, there are many types of examination in which a multiple choice test is quite inappropriate. Geometry is more concerned with method than with results; nursing is largely a practical skill; theology is a subject in which fine distinctions are important. But a combination of a test of this type with further tests in other forms may give some of the advantages of each.

The multiple choice test shares one disadvantage with every other type of examination—that it has only as much validity as its own marking scheme. Thus it is assumed that, if questions 11 and 12 each score one mark, a candidate who gets 11 right and 12 wrong is worthy of the same credit as another who gets 11 wrong and 12 right. This may sound trivial. But most examinations include questions of different types and standards of difficulty, and the choice of marking scheme then has a major effect on the order of 'merit' of the candidates.

What does an examination actually measure?

As we have noted already, different types of examination are tests of different abilities. We must look next at these various abilities in more detail, starting with those which are related to the subject of the examination.

These will vary from subject to subject, but the principal division—between tests of memory and tests of understanding—is common to most; it is not, of course, a rigid division but a finely graded one. The trend over the last few years has been away from memory work, and in some syllabuses it is specified that candidates are allowed to take reference books or lists of standard formulae into the examination room.

Also, the questions may be designed to test independence and originality of thinking, or alternatively may be more restricted in scope. They may be intended to test one aspect of the subject (grammatical correctness, say) rather than another (vocabulary, spelling, literary style, ability to present a logical argument or to understand one). Equally, the marking schemes may be designed to give credit for (or to ignore) each of these in turn.

When the examination subject is a foreign language, there are more variations than usual, including one of particular importance: that skill in translating *into* a foreign language is quite different from skill in translation from it into one's own. The distinction between the ability to speak, and the ability to write, a second language is obvious enough.

But the result of an examination depends on other things also—on the candidate's ability in matters not related to the subject being tested, and on factors outside the control of both examiner and candidate.

The fact that any written examination is a test of writing English has already been mentioned. But any form of test, including multiple choice, is also a test in *reading* English—

not a very difficult test, perhaps, but sufficiently so to cause unfairness to those for whom English is not their native language, or to those with a reading difficulty. It has been suggested that some children of immigrant families are wrongly classified as unsuitable for education in normal schools for such reasons.

Perhaps more importantly, an examination is also a test of a candidate's speed of thinking and writing, his ability to work under emotional pressure, and the experience and skill of his teacher. This last item includes skill in teaching his pupil how to pass the examination, as well as skill in imparting knowledge of the subject. There are other matters which are independent both of examiner and subject, such as emotional disturbance, physical disability, illness and, for girls, tension during the menstrual period. On all of these, and on other things besides, may depend the result of an examination and, in turn, a decision of fundamental importance in a person's career.

One other source of error is the fact that sampling is involved at various points in the examining process. Any examination paper tests only a sample—perhaps quite a small sample—of the syllabus on which it is based. If a choice of questions is allowed, then a sampling process is involved when the candidate makes his choice. (It is not, of course, a random choice, being influenced by the order of the questions on the paper and various other things; whether this makes matters better or worse must be a matter of opinion.) Also, a candidate's answers to the questions are only a sample of all the answers which he could give.

On this last point, think of a child who is given 100 sets of ten simple arithmetical problems, each set being of the same order of difficulty, so that they can be considered as 100 comparable examinations. The child will make mistakes, but not uniformly, so that some 'examinations' produce higher marks than others. Any one examination, therefore, yields a sample of his work which may or may not be a true reflection of his ability.

The element of chance, inherent in any sampling process, thus plays a major part in any examination, and it is possible that the resulting errors are as serious as any of the others which have been mentioned. When all these errors are added together, one may reasonably wonder whether reliability and validity might be so low that the whole edifice ought to be scrapped. But this radical view has one simple argument against it: inefficient and unfair though examinations may be, any alternative method of assessment is vastly worse. If assessment is necessary, so are examinations.

One improvement could be made at once. As the reader will have noted earlier, it is good statistical practice to indicate, together with any published figure, some estimate of its accuracy. A specific proposal made for the A-level examination (which could easily be adapted for others) was that there should be twenty grades instead of six, with a note on each certificate to the effect that there was a range of uncertainty of one grade either way. For example, a grade 12 would imply that the best estimate of a candidate's ability was the range of grades from 11 to 13. Perhaps because its merits were not fully appreciated, or perhaps because it was felt that the results would not be generally understood, the proposal was rejected. Either way, a little more education in statistical thinking seems to be called for.

Intelligence tests

Everything which has been said here about examinations applies equally to tests of intelligence. But in this difficult and controversial matter, there are further problems, of which two are worth a mention here.

First, there is the fact that intelligence cannot be defined or measured in any quantitative manner, independently of an intelligence test. In other words, there is no possible way of assessing the validity of an intelligence test, since comparisons with the results of other tests provide checks of reliability only. It has seriously been suggested, in fact, that intelligence should be defined as the thing which intelligence tests measure. All this does not mean that such tests are useless, by any means, but it does suggest the need for caution in interpretation.

There are numerous types in use, many of them named after the individuals who developed them—Binet, Terman's modification known as the Stanford-Binet, Wechsler, Eysenck, Cattell, Raven. These are not all designed to measure the same thing, but all yield results which tend to be labelled IQ, or 'intelligence quotient'. The various IQ's, therefore, are not all the same for any given individual. But there is general agreement that the standardisation of scales should be arranged by using a mean of 100 and a standard deviation of around 15. None of them claims to assess accurately any scores outside a range of about 60 to 140— roughly, the top and bottom $\frac{1}{2}\%$ of the whole population. Sometimes such claims (IQ's of 200 or more) are made but have little meaning.

The scale was originally based only on a mean of 100, together with the idea of a person's 'mental age'. A person with a mental age of 13 (that is, one having the intelligence level of an average person of age 13) and a chronological age of 10 was said to have an IQ of $\frac{13}{10} \times 100 = 130$. But this concept is open to objection and is rarely used nowadays.

The other principal problem is the impossibility of constructing what is known as a culture-free test. This is important in two types of situation: in comparing, individually, the intelligence levels of people of different educational, social and cultural backgrounds, and in comparing, on a larger scale, the average intelligence levels of different ethnic or regional groups.

To illustrate the difficulty, consider the following examples. In each case, the subject has to choose the odd one out of the items given, on the grounds that the other four have something in common.

1. herring   whale   shark   barracuda   cod
2. ant   spider   bee   moth   midge
3. Byron   Shelley   Keats   Chamberlain   Chaucer
4. 837   612   549   422   342

These are all to be found in a well-known series of 'intelligence' tests. But look what is required of the intelligent subject—knowledge about poets and politicians, the distinction between fishes and mammals using an example which is an exceptional case, the numbers of legs of spiders and various insects, and (if the last question is to be tackled within a reasonable time) factorisation of fairly large numbers. All four are perfectly fair questions, and the tests which include them may well give reliable results. But matters such as these are more within the scope of general educational achievement and memory than of intelligence as such.

There are, of course, many other types of test question which are much less culture-based than these, and there is not sufficient space here to justify the statement that a truly culture-free test is impossible. One can, however, state the restrictions under which such a test (if it were possible) would have to operate: it would have to exclude both the need for, and the use of, any acquired knowledge other than the basic ability to communicate. Arguably, there is no need for acquired knowledge of a few concepts such as symmetry, shapes and how they fit together, and, perhaps, ordering by size. It is more difficult to ensure that someone who *has* an acquired knowledge of such things does not gain any advantage from the fact.

When considered in terms of trying to compare the intelligence levels of the different racial groups in the United

States, which have clear cultural differences between them, the size of the problem will be seen clearly. Even with a culturally fair test, a true comparison can only be made between random samples, and the reader will be aware that this is no mere formality. It is not surprising, therefore, that claims of a 15-point difference between the IQ's of the white and Negro groups are so hotly disputed.

Finally, some examples to show that the 'intelligence' measured by an intelligence test can be a pretty obscure entity. These all come from the same source as the earlier examples and, as before, the odd one out has to be chosen in each case.

1. Jupiter Apollo Mars Neptune Mercury
2. Spain Denmark Germany France Italy
   Finland
3. August September October November
   December
4. courage leave measles steamer

No doubt of the answer to the first one—Apollo, as all the others are planets. Check with the author's own answer, and Apollo is correct. But his reason is that Apollo is the only Greek god, the others all being the names of Roman gods. Which of the two reasoning processes shows the greater 'intelligence'?

The next is more difficult. The five countries other than Finland are connected geographically, Denmark is the only monarchy, all but Spain have democratically elected governments, only Germany is (and was when the test was first published) a divided country. The choice between these is a highly subjective one, and the person whose answer is 'correct' is not necessarily the most intelligent, merely the one whose knowledge, interests and thinking processes correspond most closely with those of the person who set the question—whose answer, incidentally, was Denmark.

To the third question, the answer is so obviously August that the intelligent subject may start looking for a catch—thus putting him at a disadvantage compared with a less intelligent one, who would choose the right answer and then go straight on to the next question. But there is no catch—after careful thought I can see no reason for any other answer, although there are at least seven reasons for August.

But the last example surely deserves a medal of some kind. As already mentioned, it is from the same source as the others, and is also quoted (along with others from different writers) by Dr Alice Heim in *Intelligence and Personality* (Penguin Books). Tactfully, she allows it to remain anonymous, and so will I. As she points out, all the words except 'courage' have the same vowel pattern, all but 'leave' have two syllables (and only 'leave' can be used as a verb as well as a noun), only 'measles' is plural. But the answer given was 'steamer', for the reason that the other three are colloquially connected with nationalities—Dutch courage, French leave, German measles. Comment is superfluous.

All that can be said in conclusion, more in sorrow than in anger, is that it is a pity that such a blunt instrument ever has to be used to make fine distinctions between individuals, with consequences that are often far-reaching and where mistakes are difficult to rectify. Those who are opposed to the use of tests of this kind for choosing the type of school best suited to children of eleven years old have some powerful statistical arguments to support their case: some mentioned above, others requiring more mathematical analysis than is appropriate here. To those in a position to see the effect of 11-plus mistakes, in both directions, the argument, when brought to life in such vivid terms, looks even stronger.

14 One man, one effective vote

An election is, in part, an exercise in statistics, even though it has various features not shared by most statistical investigations. Space here does not allow a full analysis of the problems involved—how, for example, a party can gain a majority of seats on a minority vote, and the difficulties which face a party which is third in order of size. The simplest form of election is the single choice type, in which any number of candidates may offer themselves for election in a one-seat constituency, and each voter may vote for one only—and it is in this that most of the difficulties arise. Of course, the single-choice system has its defenders, and it is not intended to take sides in the matter here. But for those who do consider the single-choice system unfair, an alternative exists which seems to be growing in popularity, and it is worth including here as an instructive statistical method, quite apart from its merits as a voting system.

If an area which is fairly homogeneous in its balance of political opinion—with say 60% Conservative and 40% Labour—is divided into five single-member constituencies, all five of them, in an election based on straight majorities, would return a Conservative, and the Labour voters would remain unrepresented. The system described here is a way of removing this disadvantage, if disadvantage it is considered to be. Known as the Single Transferable Vote (STV), it was first used in a United Kingdom election in 1973 to try to solve the political problems of Northern Ireland which had arisen because a large minority of the electorate, the Catholics, were consistently under-represented in the traditional scheme. On the whole, it appeared to produce the desired result of providing non-Unionist representation; whether that will solve Northern Ireland's problems is not, of course, something in which statistics can help further. An imaginary example is the best way of showing how STV works.

The usual arrangement is for a constituency to be rather larger than in a single-choice system, and to elect a number of members between two and six. The greater the number, the better the opportunities for small parties to gain a seat, but the complexity both of organisation and of the ballot paper itself increase at the same time. In our imaginary constituency, it is assumed that there are three seats, and five candidates named Allen, Beresford, Clinton, Dawson and Evans; each voter lists all five candidates in order of preference, for example BDECA.

The number of votes cast in all is taken to be 4000—much smaller than would happen in practice, of course. As soon as this figure is known, the supervisor of the election calculates what is called a quota, as follows: the number of votes cast is divided by one more than the number of seats and, if the division comes out exactly, one is added to the result, if there is a fraction, this is ignored and one is added to the whole number part. In this case, 4000 divided by 4 is 1000, so the quota is 1001. If 4001 votes had been cast, division by 4 gives $1000\frac{1}{4}$ so that the quota would again have been 1001.

The point of this calculation is that the number thus found is the lowest number of first-choice votes which, when cast for each of the right number of candidates required to fill all the seats, would not leave sufficient votes remaining for any other candidate to gain the same number. If three candidates in this election gained exactly 1001 votes each, they would be certain of election to the three seats, because only 997 votes remain to be shared among the others.

Two procedures then follow.

First, if no candidate has reached the quota when the first-choice votes have been counted, the candidate at the bottom of the poll is eliminated, and his votes are distributed to the other candidates according to the second preferences stated.

If no candidate has then reached the quota, the next candidate up the list is eliminated in the same way, and so on.

Secondly, any candidate who reaches the quota, either on the count of first-choice votes or after the distribution of the votes of eliminated candidates, is declared elected.

But the margin by which the quota is exceeded is also important. Suppose the first-choice votes for the five candidates, in order, are 973, 786, 870, 749 and 622. No candidate has reached the quota, and so Evans, with the lowest total, is eliminated. The votes cast for him are distributed to the others according to the second choice stated by each voter; the original third choice stated on these ballot papers now becomes, effectively, a second choice, and may influence the voting at a later stage. Suppose the second preferences are, in order once again, 415, 114, 21 and 72; the totals are now 1388, 900, 891 and 821, for the four candidates still involved. Allen has now reached the quota and is declared elected.

The next stage is the most difficult, and it is helpful to write out the (imaginary) figures in full.

ABCD	255	BACD	165	CABD	169	DABC	166
ABDC	428	BADC	251	CADB	185	DACB	129
ACBD	133	BCAD	104	CBAD	129	DBAC	186
ACDB	111	BCDA	82	CBDA	88	DBCA	204
ADBC	192	BDAC	163	CDAB	176	DCAB	66
ADCB	269	BDCA	135	CDBA	144	DCBA	70
	1388		900		891		821

The votes shown as ABCD will include those who originally voted EABCD and have now been transferred according to their second choice, as well as those who voted AEBCD, ABECD and so on.

The number of first preferences credited to Allen at this

point is 387 more than the quota figure. These excess votes are distributed to Beresford, Clinton and Dawson in proportion to the second preferences stated, which are 683, 244 and 461 respectively. Thus Beresford is credited with a number of extra votes found like this:

$$683 \times \frac{387}{1388} = 190.$$

Since the 683 is made up of 255 who rate Clinton ahead of Dawson and 428 who rate Dawson above Clinton, the 190 is correspondingly made up of 71 whose next choice is Clinton and 119 for whom it is Dawson; this is not important now, but it may be at a later stage.

Beresford's total is now $900 + 190 = 1090$, and he is declared elected. Clinton's share of Allen's excess votes is 68, taking him to 959, and Dawson's is 129, giving a total of 950. A third candidate has still to be elected, and a further distribution will be necessary.

Before continuing with the calculation, it is worth noting that the total number of votes which are involved at this stage is $1090 + 959 + 950 = 2999$, which is 1001 less than the original 4000. A quota of 1001 is therefore still appropriate: if two candidates gained this number of votes and were elected to the two unfilled seats, the third candidate could not quite reach the same total. (In fact, a quota of 1000 would be sufficient; this slight discrepancy only occurs when the original division to find the quota works out exactly, and it is of little practical importance.)

The 89 votes by which Beresford has exceeded his quota must now be shared between Clinton and Dawson in proportion to the preferences next in order. But the figures in the second column of the table above are not sufficient to do this, because of the earlier distribution of Allen's surplus votes. To the 165 shown as BACD, therefore, must be added the 71 of the 190 whose next choice was Clinton, and similarly the 119 for Dawson are added to the 251. The total out of Beresford's score of 1090 who rate Clinton above Dawson is thus $236 + 104 + 82 = 422$; those who rate Dawson above Clinton, $370 + 163 + 135 = 668$. So Clinton's share, calculated as before, is 34, and Dawson's 55. Clinton's final total is 993, and Dawson's 1005. The third seat therefore goes to Dawson, in spite of the fact that he was well behind Clinton after the count of the first-choice votes. Clinton, in spite of his success at this stage, did not gain enough second and third preference votes to reach the quota.

The actual process of counting and calculation probably seems rather complicated. Indeed it is, when compared with that required for a single-choice election. The calculations are, however, easy to transfer to a computer, with automatic reading of ballot papers by a mark-sensing device. Various safeguards are, of course, required but are not difficult to arrange.

The real merits of the system become clearer when the candidates belong to certain political parties, and the voters are guided in their orders of preference by the party affiliations of the candidates. Suppose, once again, that there are 4000 voters, and six candidates for three seats; three candidates belong to the Democrat party and three to the Republicans. We assume also that 1001 voters (that is, exactly one quota) support the Democrats, and the other 2999 are Republicans.

When the first-choice votes are counted, two of the Republican candidates will reach the quota and be elected; their surplus votes will be transferred to the third Republican candidate. Since the two elected candidates account for 1001 votes each, the number finally credited to the third man will be 2999 less 2002, or 997. Two of the Democrats will now be eliminated, but their second- or third-choice votes will be credited to the third Democrat, thus bringing his total up to 1001 and securing his election.

The effect of the transfer of votes is that, as this example shows, a political party which puts forward any number of candidates for election in a constituency in which there are x seats, and can secure a fraction of the total vote at least equal to $1/(x + 1)$, is guaranteed one seat; if the fraction is $2/(x + 1)$, two seats are guaranteed, and so on. This explains why STV is often referred to as 'proportional representation', although there are other schemes designed to achieve the same result.

A further difference from the single-choice election is that a party is not at a disadvantage if it has more candidates than the number of available seats. The extra ones are eliminated, but the votes are not wasted—they are just transferred to another candidate from the same party. The effect is therefore that the party organisers do not need to choose between rival would-be candidates—the voters themselves will decide. Much the same happens if a former member of one party breaks away and stands in opposition to the official party candidate: one or other will be eliminated but, provided that the voting still follows along party lines, the opposing party will gain no advantage. In a single-choice election, such a split vote can be disastrous for the party involved.

The single transferable vote is now the system used in several countries, and in various elections of a more restricted kind such as those of trade union officials. There are some practical disadvantages, as mentioned earlier, the principal one being that a political party finds it more difficult to gain an absolute majority over two or more others. But that it gives fairer representation, and a greater opportunity for the individual voter to exercise a real influence over the election result, there can be no serious doubt.

15 Crimes, criminals and punishment

There are many reasons why the statistics of crime are un-usually difficult to record and interpret. The most obvious one is that none of those involved is able to record the figures in an informed, objective and disinterested way: the criminals themselves wish their activities to pass unnoticed, and the police can only count the number of offences known to them. The police, also, have perfectly good reasons for wanting the figures to reflect favourably on their work: this would not, perhaps, matter very much if the things being counted were susceptible to precise definition, but this is not the case.

The situation is, in fact, that there is an immense range of categories of criminal and lesser offences, often overlapping, often occurring in combination, and with definitions which may change (both by law and by custom) over a period of time. Some of these categories result from legal definitions of many years ago which have little relevance to present-day conditions. Even a simple and common word like 'burglary' can give rise to problems—it was redefined in 1968 by Act of Parliament, so that a trend going across that date can mis-lead. The difficulties with more technical terms such as 'assault' and 'indictable offence' are clear enough.

If confusion arises over the problems of interpreting, say, the statistics of housing or of unemployment, it is usually possible to appoint a team of outside specialists to investi-gate—typically, by means of a sample survey. But in matters concerning criminal activity this is very rarely possible. Sampling is difficult if not impossible, questionnaires are un-likely to produce accurate answers, and neither courts nor criminals are willing to co-operate in statistical experiments. To give just one example of what cannot be done, research into jury behaviour would yield results of immense interest, but only simulations (on a small scale and at considerable expense) have so far been attempted.

Finally, the subject is one of great public interest and, in certain matters such as murder and sexual offences, controversy also. It is always more difficult to carry out statistical analysis of a high standard in such circumstances—as, for example, when a change of definition, for which there are sound technical reasons, is liable to be delayed because it is known that a public outcry would result.

Much the same is true of penal statistics. The numbers in prison, on probation and so on are, of course, known accurately, but the *effects*, in terms of reconviction rates or the true influence on the individuals concerned, are not easy to assess. One of many reasons is that in the United Kingdom, though not in many other places, it is quite legal to live under an assumed name, so that multiple convictions of the same individual are often recorded under different names in the Criminal Records Office.

Interpretation of sentencing statistics is also far from easy. For one thing, criteria other than the nature of the crime influence both the form and the length of the sentence; further problems arise over multiple offences.

The figure which is usually of most public interest is the trend in a given type of crime over a period of time—unfortunately, a measure which is particularly sensitive to changes of definition and, in certain matters, to changing attitudes of courts, juries and police. It is often alleged, for example, that the release of restrictions on the sale of pornographic literature does not result in any increase in sexual offences, but all that may have happened is that police 'permissiveness' has changed in parallel with that of the law so that a true increase has gone unrecorded. An element of discretion, however good the reason, is often the cause of inaccurate statistics.

One particular point of definition which is important is just what counts as a single unit of a particular crime. A bomb is planted in an airliner, killing its intended victim and a hundred others; a poisoner kills eight people over a number of years; a gang of youths sets upon an old man, and four of them are convicted of murder. How many murders in each case?

The gang murder would count as one, and the poisoning as eight, on the grounds that these were the respective numbers of victims. But this approach would give a misleading result in the matter of the bomb in the airliner. Also, the concept of a 'victim' does not work well for minor property offences, fraud, arson and so on. Multiple offences, such as those involving armed robbery, also need careful classification, and the principle here is to count the most serious offence only. Further confusion may arise as a result of a 'guilty' plea being accepted for a lesser offence than that originally alleged.

For the general reader, two points may serve to summarise the essentials of the situation. First, the definitions must be such that a small number of exceptional offences (such as multiple deaths due to a bomb in an airliner) cannot have a major effect on the totals. Secondly, changes in definition and in recording practice must be stated whenever a comparison between two figures is being made. Changes in definition are inevitable in themselves, and it would be helpful if the figures for a few years prior to a change were reclassified according to the new definitions in order to make true comparisons possible. But this is not often done.

When a comparison is being made, it is essential that the same measure be used for each set of figures, quite apart from problems of definition. There are several possible methods.

The most obvious measure is that of convictions secured.

But, if a lesser charge is substituted, a record of 'no conviction' for the reported offence would be misleading. Also, it is customary in cases of multiple murder to name only one victim in the charge so that, if a conviction for this is secured, no one would be brought to trial for the others. Often, particularly for minor property offences, the defendant (when he has been found guilty of one offence) asks for others to be 'taken into consideration'—in other words, he ensures that he is immune against subsequent prosecution for them at the cost of a small increase in sentence. Such offences are counted, quite fairly, as solved.

A similar course of action is followed if there is some reason why the person clearly responsible for a particular offence cannot be brought to trial. This may be because he is under the age of criminal responsibility, because he has died or is mentally unable to understand that he has committed an offence, or because the witnesses are not able to give evidence. There is a distinction, in other words, between the kind of evidence which is required to secure a conviction and the kind which is sufficient to draw conclusions for statistical purposes.

The same procedure is followed, generally speaking, when a person is charged but acquitted. In some cases, it is, of course, true that a person acquitted has had nothing to do with the offence at all, and the person actually responsible has still to be brought to justice, but this is not universally so by any means. As a working rule, therefore, such cases are counted for statistical purposes as solved.

This gives a measure of the proportion of reported crimes which are 'cleared up'—a measure which is rather higher than the conviction rate and, in general terms, a more realistic one. Rightly or wrongly, it is often taken as an indicator of police efficiency, particularly in the investigation of crimes of violence. But, in common with all statistics concerned with crime, it should be treated with caution.

16 Demographic statistics

No thinking person nowadays is unaware of the problems being created by the rapid increase in the world's population. The study of population statistics is known as demography, and in many ways it is not a difficult subject for the layman to understand: the quantities involved are countable and to some extent easy to visualise, and most of the factors involved are within the ordinary person's experience. So, for example, one can illustrate the situation by saying that in England the increase each year is equivalent to the population of a medium-sized town, or that we need to build one secondary school every two days to keep pace with the bigger numbers. In analytical terms the simplest measure of population change—the percentage annual increase—is not a difficult concept, and comparisons between different countries are quite straightforward.

But when one tries to assess the causes of change numerically, matters are very different. It is fairly obvious, for example, that if contraceptive methods become more widely used the resulting change in the birth rate will slow down or perhaps reverse the present rate of increase in the population. The precise results of a particular change in the birth rate are, however, much less obvious.

When both birth and death rates are changing together, as they usually are, the problem becomes complex indeed, even before migration is brought into the picture. Various attempts have been made to simplify matters, but none has proved wholly satisfactory; the advent of the computer has made major simplification unnecessary anyhow. But, even with the help of computers, population projections have often proved so inaccurate that there is a widespread reluctance to take them seriously. (Demographers, perhaps somewhat defensively, prefer to speak of 'projections' rather than 'forecasts'.) However, as the figures have been criticised more often for appearing too high than appearing too low, it should be said that projections have proved too low far more often than the reverse; almost the only exceptions to this have been for individual industrial countries.

The first stage in developing an understanding of population changes is to be able to describe an existing population. The distribution by age and sex is known with reasonable accuracy for most countries, and may conveniently be shown diagrammatically in a form known as an age pyramid. Low birth rates combined with low death rates tend to produce a thick column tapering off at the top; high rates of both make for a wide base and a thinner column.

Next we require the pattern of death rates. This can be shown graphically as a survival curve, which gives the number still alive out of a nominal 1000 live births at each point during the next 90 or so years. The graph shows an initial drop on account of infant mortality, which is defined as the

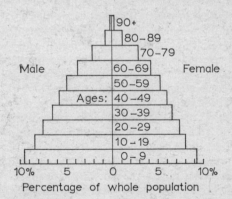

Figure 21 Age pyramid for low-mortality population with a relatively low growth rate

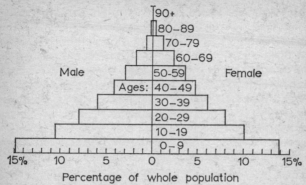

Figure 22 Age pyramid for high-mortality population with a relatively high growth rate

number per 1000 live births who die within their first year, and then levels off.

For life assurance purposes, death rates are given as the proportion of a particular age group who may be expected to die within the next 12 months (or other specified period). Thus if 625 out of the original 1000 are alive at age 65 and,

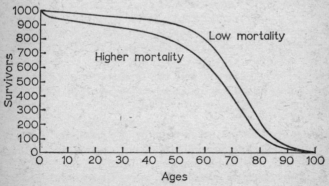

Figure 23 Survival curves

of these, 25 are expected to die within 12 months, the death rate is 25 in 625, or 0·04. If 30 of the remaining 600 are expected to die within the following 12 months, the rate is then $30/600 = 0·05$. Note that the 2-year rate would not be 0·09, because the proportion for the second year is related to a base of 625 rather than 600; the actual figure is $55/625 = 0·088$. There is also an awkward discontinuity where the rate changes from 0·04 to 0·05, whereas the change in the true rate is a smooth one.

These discrepancies can be avoided by using a rate which is related, not to the number alive at the start of the relevant period, but to the number alive at any given time. This is called the age-specific mortality. But the life assurance method is easier to understand, and is quite adequate for most purposes.

Death rates are quoted separately for men and women; the figures differ quite considerably, as mentioned below.

A further method of expressing mortality is to use the idea of an expectation of life. This is particularly useful in giving a single figure to represent the average length of life which a new-born baby may expect to have—the expectation of life at birth. It is also used in life assurance to give the average number of years remaining to a group of people of specified age and sex, and where appropriate also of a particular occupation. One point of interest is that, if you start life with an expectation of, say, 70 years, your expectation when you reach 30 will be more than 40 years: this is because a few of your peers will have died before that time, so that you are allowed an extra year or two to make the average turn out right.

Quite apart from this, expectation of life is increasing in most countries: in Britain that at birth has recently been going up by about one year every ten years. The rate of increase was much higher earlier in this century, and there are now signs of levelling off. Women in general live longer: on current rates about six years longer than men, and the gap is still widening.

The changes in mortality mean that the measure is somewhat unrealistic, because it has to be based on current rates and, in seventy years' time when this year's babies are reaching the ends of their allotted spans, the values may be very different.

Mortality rates may also be quoted per 1000 total population, but this is not usually very helpful because the figure depends too much on the age structure of the population.

Statistics of births are covered under the general title of fertility. This comprises the number of children born to each woman, the age of mothers at the first birth and the intervals between successive children. Some of the changes in these may result from changes in the marriage pattern, including remarriage, as well as illegitimacy, contraception and abortion. The birth rate may be quoted in various ways: per 1000 population, per 1000 female population and per 1000 female population of child-bearing age. All of these are inadequate on their own for one reason or another and, to enable comparisons to be made between different countries and different periods of time, it is really necessary to give fairly complete figures for the numbers of children being born to women of each relevant age. Even then, problems may arise in the event of changes in the pattern because, if women now in their twenties produce more babies than this age group used to do, they may produce correspondingly fewer in their thirties than those who are at present in their thirties.

Births are not equally divided between the sexes: the proportion varies slightly, but is about 48·5% girls and 51·5% boys. The imbalance is somewhat greater in first babies.

The complete pattern of both fertility and mortality is required in order to assess fully the trend in the population. Since the pattern is so complicated, attempts have been made to simplify it by using a single measure or group of measures.

The best-known of these is the net reproduction rate or NRR, but neither this nor any other measure really gets over the difficulties caused by variations in the age structure of the population. The NRR is useful only as long as this limitation is recognised: it is defined as the average number of daughters who will be born to a girl who is now a baby, if current fertility and mortality rates remain unchanged. Under these conditions, a population will exactly replace itself when the NRR is 1·00. But the rates do not in practice remain unchanged, either from year to year or over longer periods of time. The short-term changes which result from, say, the absence of men at a war and their subsequent return have a considerable influence on the NRR but much less on the percentage rate of population increase. Some examples to illustrate the longer-term effects of changes are included later in the chapter.

The NRR is also independent of all mortality rates other than those for women up to the end of child-bearing age, so that changes in mortality (and consequently changes in population) can occur while the NRR remains at 1·00. The implication that a net reproduction rate of unity is the same thing as zero population growth is thus quite false, except under wholly stable conditions.

In addition to fertility and mortality, the amounts of immigration and emigration clearly have an effect on population. Not only the numbers but also the age and sex structure of the migrant groups are important, and full information on this subject is not usually available, particularly where national boundaries are uncontrolled. So, when migrants form a high proportion of the total annual changes, population projections are correspondingly less accurate than if they were based on a country with closed frontiers. In the examples which follow, migration is not taken into account.

If we are to look at the whole pattern of population

change, it is necessary to simplify if the result is to be of manageable proportions. The aim will be to clarify general principles rather than to construct detailed projections, and changes which are either small or short-term will not be considered. No figures relating to a particular country are used, though the chosen fertility and mortality rates are fairly typical of the various kinds of country mentioned.

National populations are currently increasing at rates between about zero and 4% per year, with a mean of about 2%. The patterns of growth fall mostly into one or other of two categories, which are called here 'industrial' and 'developing'. A third category is made up of countries in the process of changing from the second type to the first. It is characteristic of an industrial country that it has low rates of both fertility and mortality, and of a developing one that it has high rates of both; in the transitional stage a high birth rate may be accompanied by a low death rate.

The first example is an industrial country in which the growth rate is 1% and has been stable for many years. The table below sets out the assumptions made about the population as it was in 1970, with one further column giving the age pyramid ten years later.

Age	Fertility	Mortality	Survivals	AP 1970	AP 1980
0		3%	1000	1 872	2 067
10		1%	970	1 644	1 816
20	1·4	1·5%	960	1 474	1 628
30	1·1	2%	946	1 315	1 452
40	0·2	4%	927	1 167	1 289
50		12%	890	1 013	1 120
60		32%	783	807	891
70		60%	533	496	549
80		80%	213	180	198
90		100%	43	32	36
				10 000	11 046

The principal form of simplification is to do everything in 10-year steps. It will be found helpful to take the date given to be the centre-point of each 10-year period, so that 1980 implies 1975 to 1984 inclusive, and those in the age group labelled 20 are, in fact, in the range 15 to 24 at the centre-point of the period. This may seem to cause some difficulty with the bottom age group, but age -3 in 1970 implies the babies due to be born in 1973. It is possible to pick all sorts of holes in the methods needed to allow for this form of simplification—partly because of the methods themselves, partly (on a more technical level) because the process is deterministic rather than stochastic, meaning that it takes the quoted rates as fixed rather than as mean values of a range of figures. But for seeking out principles they are quite good enough, and the results will be found fairly consistent with what happens in practice.

It is also assumed that births are equally divided between the sexes, and that mortality rates are the same for each sex. The assumed fertility rates in the table mean that each woman produces on average 1·4 babies during the 10-year period starting at age 15, and so on. The number of women in each group is half the total population in the group, and of their babies half will be girls. The mortality rates are for a 10-year period: that is, 32% of those who survive to age 60 will die before they are 70, and so on. Once again, some approximations are inherent in the method.

From the mortality rates, survival figures can be calculated, starting with a nominal 1000. Out of this number 3% will die before they reach 10, leaving 970, another 1% of the 970 before age 20, and so on until the 43 survivors to age 90 suffer 100% mortality. (This ignores the centenarians, but they are few in number.)

The 1970 age pyramid has been calculated on the assumption, mentioned earlier, that the rate of population increase has been constant at 1% for many years. Over a 10-year period, therefore, each age group is 10 % larger than the one it replaces—or rather, just over 10 % because the increase is exponential. The method of calculation will not be described here, though anyone who follows the rest of the chapter fairly closely will be able to work it out for himself. The total 1970 population has been arranged to be a nominal 10 000, which can be scaled up as desired to suit a country of any given size.

The 1980 population may now be worked out. Out of the 1872 people given age 0, 3 % (56) will die leaving 1816, 1 % of 1644 will die leaving 1628 and so on. The number of babies is found as follows. Of the new population aged 15–24, half will be women and these will produce an average of 1·4 babies each, a total of $1628 \times \frac{1}{2} \times 1·4 = 1140$. Similarly, the 25–34 age group will produce $1452 \times \frac{1}{2} \times 1·1 = 799$, and the 35–44 group $1289 \times \frac{1}{2} \times 0·2 = 129$. The total of these is 2068. The tabulated values have been calculated with greater accuracy than that used here, and the correct value is 2067, as shown in the table.

The total population in 1980 can now be found by addition, and the increase is about 10·5 % over the 1970 figure, both in the total and in each separate age group.

Various other measures of population structure and change may be found from these figures. The number of female babies born to each woman who survives to age 45 is half the sum of the fertility rates, or 1·35; this is the *gross* reproduction rate, which is not a very useful measure on its own. The net rate or NRR is rather smaller, because it takes account of mortality; thus the effective fertility of the 15–24 group is based on the survival number of 970, not the full 1000. A calculation along these lines shows that the NRR is 1·285—as expected, rather more than 1 because the population is growing in size under uniform conditions of fertility and mortality. The birth rate per 1000 population is 18·7.

The mean age at child-bearing is 25·5 years, which is a

fairly typical figure. This is, of course, a measure of the length of one generation.

The mean age at death, which is the same thing as the expectation of life at birth, is 67·6 years—once again, a roughly typical figure for an industrial country. (Anyone wishing to check this figure should note that a continuity correction of half a decade is required: those who die in their sixth decade have an average life of about 55 years, not 50.)

One further measure is useful: the proportion of dependants to producers. The 'producers' are taken to be those in the age group from 15 to 59, assuming that their earning power in the community provides for the young and the old as well as for themselves. Arguably, both age limits should be rather higher in an industrial country, but these values are chosen to facilitate comparisons between different countries. The higher the number of dependants per producer, the greater will be the country's economic task. The value here is 0·73.

The growth pattern can now be continued as long as desired, and in this particular case the growth rate remains constant at 1%, doubling the population every 70 years or so, and the age structure remains stable. There is nothing difficult about the calculations and, although those that follow have been done by computer to save labour, there is in principle nothing that a few hours' work with a slide rule could not do just as well. But, with a computer to do the hard work, it becomes possible to alter some of the assumptions to see what happens, and this is where the results start to look interesting.

Suppose that in 1965 our imaginary industrial country decides (we need not concern ourselves with motives or methods) that the population has reached its desirable limit, and that the birth rate should be cut. The result of this is, say, that the fertility rates for 1965–74 change from 1·4/1·1/0·2 to

1·15/0·8/0·15: this cuts the NRR down from 1·285 to 1·000. Mortality rates are unaltered. The population totals then work out as follows, starting with a nominal 100 in 1970. The figures are rounded and so not directly checkable against each other.

Year	1970	1980	1990	2000	2010	2020	2030	2040	2050
Population	100	106	113	117	121	124	125	126	126
% inc. p.a.		0·59	0·63	0·43	0·32	0·22	0·11	0·03	0·00

This demonstrates the point made earlier, that a NRR of 1·00 is not the same as a zero growth rate. The annual rate drops from its previous value of 1% to about 0·6% at once, but it stays there for two decades, and does not fall to zero for about seven decades. During that time the population grows by about 25%.

Since this assumes stable mortality, it underestimates the growth which would occur in practice because mortality rates are generally falling, for all except the highest age groups. Also, a 22% fall in fertility over such a short time is most unlikely, and if spread out over a longer period the time to stability would be correspondingly longer. If the fall is spread over three decades, a further five decades are required before the population settles down, and the stable population after this time is about 41% above the 1970 figure instead of 26%.

The reason why the growth rate does not fall more quickly can be seen in the full figures. With the original fertility rates, the increase in population is 1046 per 10 000 during the first decade, of which only 195 are accounted for by the increase in the lowest age group. The remainder result from the fact that every age group is about 10% bigger than the one it replaces. When births drop by 22%, there are 454 fewer babies, and hence a fall of 259 in the lowest age group instead of the 195 rise. But the other increases continue as before, since all those aged 10 or more were already born when the change in birth rates took place. And it takes more or less a full life span before the increase of births in 1965–74 compared with 1955–64 works its way out of the system.

The next example is a 'developing' country. The characteristics here are high fertility combined with high mortality, in contrast to the industrial type in which both are low. The rates can reach much higher levels than those shown here— infant mortality rates up to 300 per thousand and, very rarely, over 500 per thousand, combined with gross reproduction rates up to 3 or 3·5. As before, we assume a stable balance within the population to start with.

Age	Fertility	Mortality	Survivals	AP 1970	AP 1980
0		8%	1000	2 898	3 707
10		2%	920	2 085	2 666
20	2·12	3%	902	1 599	2 043
30	1·72	5%	875	1 210	1 551
40	0·36	8%	831	899	1 150
50		18%	764	647	827
60		40%	627	415	531
70		70%	376	194	249
80		85%	113	47	58
90		100%	17	6	7
				10 000	12 789

The rate of increase is 28% in 10 years, which is 2·5% per year. The gross reproduction rate is 2·1, the NRR 1·86, and the expectation of life at birth about 59 years. The birth rate per 1000 is 29, the length of a generation 25·7 years, and the dependants-producers ratio 0·85. None of these is an extreme value by any means.

Apart from the more rapid growth rate, the only difference from the industrial country which requires comment is the increased number of dependants per producer. The figures here should not be taken as typical, because values can vary widely in non-stable populations, but it is generally true that the ratio is greater in the developing type.

The next population we shall look at is that of a developing country which changes gradually, over a 50-year period, to an industrial type. This implies a fairly rapid change, but not exceptionally so. The change is brought about by altering the fertility and mortality rates in five equal steps, from the figures above to approximately those used earlier for the industrial population with a NRR of 1·00; the final mortality rates are slightly lower, to allow for the fact that the rates for industrial-type populations are still going down.

Age	Fertility	Mortality	AP 1970
0		8% → 3%	2 898
10		2% → 1%	2 085
20	2·12 → 1·15	3% → 1·5%	1 599
30	1·72 → 0·8	5% → 2%	1 210
40	0·36 → 0·15	8% → 3%	899
50		18% → 10·5%	647
60		40% → 30%	415
70		70% → 55%	194
80		85% → 75%	47
90		100%	6
			10 000

As in the earlier example which used a decreasing birth rate, the rate of increase in the total population is much slower to react than the NRR itself. In fact, it does not come down to near zero until 70 years or so after the end of the

50-year transitional period, as shown by the figures below.
As before, the population total is related to a base of 100 in
1970.

Year	1970	1980	1990	2000	2010	2020	2030	2040	2050
Population	100	128	160	196	234	271	302	330	351
% inc. p.a.	2·50	2·26	2·08	1·79	1·46	1·08	0·92	0·63	
NRR	1·86	1·69	1·52	1·35	1·18	1·00	1·00	1·00	

Year	2060	2070	2080	2090	2100
Population	365	374	377	378	378
% inc. p.a.	0·41	0·24	0·10	0·02	0·00
NRR	1·00	1·00	1·00	1·00	1·00

The figures for the dependants-producers ratio are not
given, because they are too sensitive to the balance between
the assumed fertility and mortality rates to give a fair impression. Otherwise, however, the picture is a clear one. The
momentum of the relatively high rate of increase in 1970–80
cannot be stopped as quickly as the NRR changes, and the
annual growth rate is still over 1% when the NRR has
reached the 'replacement' level of 1·00. After this point there
is a rise of almost 40% before stability is reached; the total
increases by a factor of 3·7 over the 100-year period to 2070.

As a final example, consider a modified form of that above,
in which mortality rates fall while fertility remains constant.
The annual growth rate now increases from its already high
value of 2·5% to 2·8%, so that the population doubles itself
every twenty-five years, and would increase by a factor of 16
over a century unless the birth rate were brought down. This
is the kind of growth which spells economic disaster, as the
working members of the community struggle under the
double burden of providing for the increasing numbers of
children and creating the capital required for industrial,

agricultural and social development. If, in the face of the struggle to keep alive, capital investment is deferred, the increase in education which is usually followed by (if not actually the cause of) a lower birth rate is also deferred, and the growth rate which is the root of the whole problem remains high. Some Latin American countries have already got into this situation; others in Africa may soon find themselves in it as well.

*　　*　　*

It should be clear to the reader at this point that the problem of making population projections divides itself into three parts.

First, we have to understand the current situation—the size of the population, its age structure, fertility and mortality rates, and migration. In most developed countries, accurate information on these matters is publicly available, with two exceptions: migration and fertility. Migration has been mentioned already; the lack of information on fertility concerns not so much the total numbers of births as their classification, by age of mothers particularly. (This information has only been required to be registered in Britain since 1960.) In other countries, the published information is sometimes much less: the great majority have a census of some kind, but not all have any registration system for births and deaths, and in some cases there are estimates given by different organisations for the whole population which differ by a factor of 2 or more. The totals given by the UN Statistical Office, incidentally, are based on the assumption that the overestimates in some cases will cancel out the underestimates in others: it is very doubtful whether this is true, though until better figures are available there is no other way round the difficulty. Because of this wide range of quality in the data, it is not easy to give a fair summary of the whole situation. But a personal impression is that sufficient good-quality information is available, for a wide variety of individual countries, to confirm the general conclusions which may be drawn from the (less accurate) world totals.

Next, we have to estimate the changes in the rates of births, deaths and migration over the whole period for which the projection is required, and this is where the guesswork really starts. Migration, of course, is very difficult to predict: it may be governed by law, but over twenty or forty years the laws may change. As far as the world total is concerned, however, this is one case where the errors for individual countries do cancel each other out. Two methods are available for estimating changes in fertility: one is simply to watch the trend, and the other is to conduct a survey of women of the relevant ages to find out their future intentions on starting or increasing their families. Needless to say, all sorts of things can go wrong with either method: intentions are not the same thing as facts, and trends can alter for various reasons. Possible future developments which would cause changes are a self-administered abortion method, a 'morning-after' contraceptive and a once-per-month contraceptive pill whose use could be medically supervised, for use in countries with low educational standards.

The final stage is to calculate the changes in population which will result from these assumptions. This involves no further guesswork and no significant approximations. Too often the criticisms directed, rightly or wrongly, at population projections concentrate on this stage of the work, which is a purely arithmetical exercise checkable by anyone who has the necessary technical skill. It is the assumptions (particularly on fertility) which should be the target of criticism. The Registrar General's projections state these very carefully, but the values assumed are very rarely stated in press reports, let alone criticised, while criticisms of the projections themselves are frequent. It is not a subject which is well covered by the press, even by the 'quality' papers.

Various points of definition and interpretation may conveniently be mentioned here. To start with, the term 'population' itself must be defined with some care. The usual way is to count the actual numbers within the frontiers at an agreed point in time, which gives the 'de facto' or 'enumerated' population. Holidaymakers and business travellers, in both directions, therefore cause errors, and so a census is taken at a time of year away from the main tourist seasons. If people are counted as if they were at their normal homes, so that visitors are omitted and absentees included, the figure obtained will be the 'resident' population. Thirdly, if members of the armed forces stationed abroad are included, together with similar civilian categories, and corresponding foreign groups excluded, the result will be the 'total' or 'de jure' population. Ideally, the definitions of the limits should be chosen to suit a particular investigation, but usually only one total is conveniently available and has to be used even if not ideal. Most countries adopt the 'de facto' definition, following the United Nations recommendation.

Some general comments on the accuracy of population data have been noted above, but a few particular points should also be mentioned. Numbers of births are usually recorded by date of registration rather than by the actual date of birth, and the interval between the two is far from consistent—it may be as long as twenty years. The need to qualify for some form of benefit may produce a flood of registrations, and the fact that registration may result in liability for tax also has its effect. Even where governments are aware of the need for accurate statistics, minority tribal groups and inhabitants of remote areas are difficult to enumerate.

Mortality figures can be misleading when quoted for unbalanced groups: rates for an area to which many people retire may appear unreasonably high, and rates for dangerous occupations such as soldiering in wartime may appear low, simply because the number dying from natural causes is much smaller than the average over the whole population. It is possible to standardise the figures for mortality in such circumstances, in a way that allows for the differing age structures, so that valid comparisons may be made.

Two other technical terms are used. One is nuptiality, meaning the pattern of marriages including remarriage, and also separation, divorce and widowhood. The other is fecundity, meaning the physical ability of a woman to re-produce. A fecund woman, married and taking no effective contraceptive measures, normally conceives, so that fecund-ity is a measure of the maximum possible fertility. Accurate statistical information about it is for practical purposes impossible to obtain.

Finally, a brief look at the whole population pattern. At the time of writing the world rate of increase is about 2·0% per annum, which means that the population doubles itself every thirty-five years. The rate of increase has itself been increasing fairly steadily. In medieval times it was not much over a tenth of the 1970 rate; by the beginning of the Indus-trial Revolution the figure was about a quarter of the same rate. (If it had stayed there, incidentally, the world popula-tion now would be a little less than half what it actually is.) Even by 1925 the rate was only 1%, half of the present rate; it is the last fifty years which have seen the most rapid changes in the pattern.

The simplest form of population projection is one based on the current growth figure. This is not very accurate, but at least it serves to make crystal-clear that, unless the pattern changes quite considerably, the situation must reach crisis point. By continuing the exponential curve at the same rate one can conjure up absurd pictures, such as that suggested in a United Nations report which pointed out that in 600 years' time there would be only 1 square metre of solid ground for each person to live on. The inevitability of crisis is beyond dispute, without a change in the pattern; the only questions are how and when, or, if the crisis *is* to be avoided, how the required changes are to be brought about.

Further, there are reasons for thinking (even without the kind of detailed analysis which was mentioned earlier) that the probable decrease in fertility rates may be partly offset by decreases in mortality—perhaps even exceeded. The main factors involved in mortality changes are the improvements in medical care, public health, housing and social conditions that all nations rightly aim towards. No one would argue that the infant mortality rate in a developing country should be kept at 10% rather than be brought down to the 2% which is typical of more advanced nations. Equally, no one would think of cancer or heart disease as a valuable agent in keeping the number of old people down to manageable proportions. Nevertheless, lower mortality rates mean that fertility rates also have to be brought down simply to keep the annual growth rate at its present level. This is why the growth rate is now at the highest level it has ever been in human history—mortality rates have steadily improved, and the drop in fertility has not been sufficient to keep pace.

The reasons for the decrease in fertility are well known, at least in their total effect. Abortion rates are generally increasing, and likewise the use of contraceptives—even in countries where officially a religious prohibition is in force. In some countries (notably India) Government action has been taken to discourage childbirth, both by contraception and sterilization. Improved educational standards tend to produce later marriages and thus fewer children. But the total effect of these changes in the last twenty years or so has been small in national terms and negligible in global terms.

So, if there is no major change in the fertility pattern—and it seems clear that this can only be brought about by strong Government action, not just in one or two countries but far more widely—it is only a matter of time before the world population growth is restrained by catastrophic action.

This was foreseen by Malthus, well-known as a pioneer in demography, in 1798: he suggested that, while populations increase exponentially, the means of subsistence can only increase at a uniform rate and must eventually fall short of requirements. At this point 'positive checks' on the population would take place. His ideas have been faulted in various ways, and his figures are even roughly accurate only for a pre-industrial society. But in spite of the errors, the principle, expressed in the most general terms, still holds. Possible 'positive checks' are wars, epidemics, famines and natural disasters. Epidemics on anything like the scale of the Black Death, in which over a quarter of Europe's population died, are unlikely nowadays, and those on a smaller scale will have little total effect. Similarly, natural disasters: a tidal wave killing a million is a major disaster in human terms, but statistically it is of little global importance. The choice lies between wars and famines.

It is not a pleasant thought. But for those who (with the highest possible motives) suggest that it would be wrong to limit people's freedom to produce as many children as they want, it is a reality which must be faced. To produce even a fairly small change—such as keeping the world total down to 6000 million by the end of this century instead of the figure of almost 7000 million currently predicted—requires action very soon indeed.

Index